New Scientific Applications of Geometry and Topology

Recent Titles in This Series

(Continued in the back of this publication)

AMS SHORT COURSE LECTURE NOTES
Introductory Survey Lectures
published as a subseries of
Proceedings of Symposia in Applied Mathematics

Proceedings of Symposia in
APPLIED MATHEMATICS

Volume 45

New Scientific Applications
of Geometry and Topology

De Witt L. Sumners, Editor

Nicholas R. Cozzarelli
Louis H. Kauffman
Jonathan Simon
De Witt L. Sumners
James H. White
Stuart G. Whittington

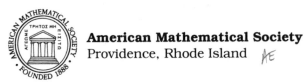

American Mathematical Society
Providence, Rhode Island   AE

LECTURE NOTES PREPARED FOR THE
AMERICAN MATHEMATICAL SOCIETY SHORT COURSE
NEW SCIENTIFIC APPLICATIONS OF GEOMETRY AND TOPOLOGY
HELD IN BALTIMORE, MARYLAND
JANUARY 6–7, 1992

The AMS Short Course Series is sponsored by the Society's Program Committee on National Meetings. The Series is under the direction of the Short Course Subcommittee of the Program Committee for National Meetings.

Library of Congress Cataloging-in-Publication Data

New scientific applications of geometry and topology / De Witt L. Sumners, editor; Nicholas R. Cozzarelli ... [et al.].
        p.   cm. — (Proceedings of symposia in applied mathematics, ISSN 0160-7634; v. 45. AMS short course lecture notes)
    The short course was held in Baltimore, Md., January 6–7, 1992.
    Includes bibliographical references and index.
    ISBN 0-8218-5502-6 (acid-free paper)
    1. Geometry, Differential–Congresses. 2. Knot theory–Congresses. 3. Science–Mathematics–Congresses. I.   Sumners, De Witt L. II.   Cozzarelli, Nicholas R. III.   American Mathematical Society. IV.   Series: Proceedings of symposia in applied mathematics; v. 45. V. Series: Proceedings of symposia in applied mathematics. AMS short course lecture notes.
QA641.N42   1992   92-26335
516.3'6–dc20   CIP

1991 *Mathematics Subject Classification.*
Primary 53A05, 57M25.
Secondary 82B20, 82B41, 82D60, 92C40, 92E10.
Copyright © 1992 by the American Mathematical Society. All rights reserved.
Printed in the United States of America.
The paper used in this book is acid-free and falls within the guidelines
established to ensure permanence and durability. ∞

Portions of this volume were printed directly from author-prepared copy.
Portions of this volume were typset by the authors using $\mathcal{AMS}$-TEX,
the American Mathematical Society's TEX macro system.

10 9 8 7 6 5 4 3 2 1   97 96 95 94 93 92

Table of Contents

Preface

Geometry and topology are subjects generally considered to be "pure" mathematics. Both originated in the effort to describe and quantize shape and form in order to understand the "real" world. Both enjoy a robust and sustained internal intellectual life, abstracted from the "reality" of their origins. Recently, some of the methods and results of geometry and topology have found new utility in both wet-lab and theoretical science. Conversely, science is influencing mathematics, from posing questions which call for the construction of mathematical models to the importation of theoretical methods of attack on long-standing problems of mathematical interest.

A case in point is the subject of knot theory, which is utilized to a greater or lesser degree in each of the six papers in this volume. Knot theory traces its mathematical origins to the work of Gauss on computing inductance of linked circular wires, and to the work of Kelvin and Tait on the vortex theory of atoms. Knot theory is the study of entanglement and symmetry of elastic graphs in 3-space. It has proven to be fundamental as a laboratory for the development of algebraic topology invariants and in the understanding of the topology of 3-manifolds. During the last decade, laboratory scientists have become increasingly aware that the analytical techniques of geometry and topology can be used in the interpretation and design of experiments. Chemists have long been interested in developing techniques that will allow them to synthesize molecules with interesting 3-dimensional structure (knots and links). Polymer scientists study the chemical and physical ramifications of random topological entanglement in large molecules. Models for molecular structure must be built and understood; reactions which produce specific 3-dimensional shapes must be designed; chemical proof of structure must be produced, and these proofs often involve the use of topology to interpret data such as NMR (Nuclear Magnetic Resonance) spectra. Molecular biologists know that the spatial conformation of DNA and the proteins which act on DNA is vital to their biological function; moreover, differential geometry and knot theory can be used to describe and quantize the 3-dimensional structure of DNA and protein-DNA complexes. Biologists devise experiments on circular DNA which elucidate 3-D molecular conformation (helical twist, supercoiling, etc.) and the action of various important life-sustaining enzymes (topoisomerases and recombinases). These experiments are often performed on circular DNA molecules, in which changes in the geometric (supercoiling) or topological (knotting and linking) state of the DNA can be directly observed–witness the beautiful electron micrographs of DNA knots and links. The recently acquired ability to preform these experiments provides a challenge for mathematics–to build mathematical models for DNA structure and enzyme action which explain the experimental observations (both qualitatively and quantitatively), models which can be used to design further experiments and to predict the outcome of these experiments. Knot theory has also been involved in a recent fundamental and revolutionary theoretical development, where the pioneering work of Vaughn F. R. Jones has set off an explosion of interaction between theoretical physics and mathematics. A whole new spectrum of invariants for 3-manifolds has been born. These new interactions between science and mathematics form a beautiful concrete example of, in the words of E. E. David, Jr, *the seemingly inevitable utility of mathematics conceived symbolically without reference to the real world* ".

I would like to thank Carole Kohanski and Jim Maxwell for their help in organizing this short course at the 1992 Baltimore AMS meeting, and to Donna Harmon for her help in the preparation of this volume. On behalf of the American Mathematical Society, I would like to acknowledge a grant from Genentech, Inc., sponsors of this AMS Short Course. Thanks also to the enthusiastic audience who attended the lectures; their interest, questions and comments were stimulating to all!

De Witt L. Sumners

Proceedings of Symposia in Applied Mathematics
Volume 45, 1992

Evolution of DNA Topology:
Implications for Its Biological Roles

NICHOLAS R. COZZARELLI

1. Scope of the Article

I shall write about DNA topology in this article in a way that I have not done before, but in the manner in which I often teach the topic. I shall provide a "teleological" interpretation of DNA topology: i.e., why and how the topological structure of DNA evolved. All biologists are fascinated by evolutionary explanations because they seek to answer the fundamental question: Why are we the way we are? Evolutionary explanations are also heuristic in that they often provide the hunch that is behind the stated rationale for an experiment. Despite this, teleological arguments are sometimes viewed as unscientific and thus are rarely written down. It is extremely difficult to determine how something evolved, let alone why, especially for something so basic to life as the genetic material. The reader is forewarned that most of what I shall write is unproven. I do, however, invite the reader to share in these secret pleasures of biologists.

2. Topological properties of DNA

A schematic representation of a small section of double-stranded DNA (deoxyribonucleic acid) is shown in Fig. 1. The DNA is depicted as a helically twisted ladder. The uprights are polymers of the sugar, deoxyribose, esterified to phosphoric acid and represent the backbones of the molecule. The rungs are pairs of four specific organic bases. It is the sequence of the bases that encodes the genetic information in DNA. A single strand of DNA consists of one backbone plus the adjacent bases. The two strands of DNA come apart during the duplication of genetic material by breakage of the weak bonds between the base pairs. Each single strand thus acts as a template for the synthesis of the other (complementary) strand. DNA is called a nucleic acid because it is found in the compartment of the cell called a nucleus

1991 Mathematics Subject Classification: Primary 92C40.57M25.

The research on which this summary is based was supported by grants from the National Institutes of Health and the National Science Foundation.

This paper is in final form and no version of it will be submitted for publication elsewhere.

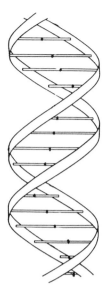

Fig. 1. Twisted ladder model of duplex DNA. The helically twisted uprights of the ladder represent the backbone of DNA, which is a polymer of deoxyribose esterified to phosphate. The rungs represent the base pairs whose sequence specifies the genetic message.

and it contains phosphoric acid. RNA (ribonucleic acid) is a related molecule in which ribose replaces deoxyribose. It generally has only one strand.

DNA inside of cells is a very long molecule with a remarkably complex topology. Topological properties of DNA are defined as those that can be changed only by breakage and reunion of the backbone. There are three important topological properties of DNA: the linking number between the strands of the double helix (Fig. 2), the interlocking of separate DNA rings into what are called catenanes (Fig. 3A), and knotting (Fig. 3B). The linking number of DNA in all organisms is less than the energetically most stable value in unconstrained (relaxed) DNA. This puts the DNA under mechanical stress which causes it to buckle and coil in a regular way called (–) supercoiling (Fig. 4). The (–) sign indicates that the linking number is less than that in the relaxed state. The name **super**coiling arises because it is the coiling of a molecule which is itself formed by the coiling of two strands about each other. Although supercoiling is, strictly speaking, a geometric property, it is a consequence of a topological one, the linking number difference between supercoiled and relaxed DNA.

Although all three topological properties are strictly defined only for DNA rings, they are important descriptors of virtually all DNA in cells, most of which is linear. This is because cellular DNA is usually so long that it is subdivided into topologically constrained loops of roughly 100 kilobase pairs (kb), which can be supercoiled, catenated, and knotted. (The size of DNA is most usefully expressed in terms of the number of base pairs rather than in the traditional length or mass units, because it directly gives the amount of genetic information.) The domains are maintained by binding of the DNA at intervals to a protein scaffold. To set a scale for the size of a topological domain, the total length of DNA in a human egg or sperm equals about 3 million kb. In addition, cells may contain small DNA molecules in the range of 5-50 kb, which are often intact closed rings.

Fig. 2. The linking number of DNA. The winding of the two strands of the DNA double helix about each other, represented by filled and open tubes, can be measured by the linking number between the strands. It is equal to one-half the number of signed crossings of the two strands in any projection of the molecule; the sign convention is given in Fig. 3. The linking number of the small circular DNA in the diagram is equal to 12, but it is orders of magnitude larger for naturally occurring DNA.

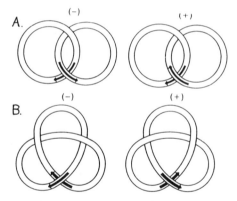

Fig. 3. DNA catenanes and knots. The duplex DNA is depicted as a tube; the arrows indicate the orientation of the DNA as defined by base sequence. The simplest examples of catenanes (A) and knots (B) are shown. There are two crossings in the case of the singly linked catenanes in A and three for the trefoil knots in B. The topological sign of the crossings is given by the convention implicit in the drawings. A (–) crossing is one in which a clockwise rotation (<180°) of the arrow on top is needed to make it congruent to the underlying one; a (+) crossing is one in which a counterclockwise motion is needed for the operation. The two possible topological isomers of each form are shown. As the number of crossings in a knot or catenane increases, the number of possible isomers grows exponentially.

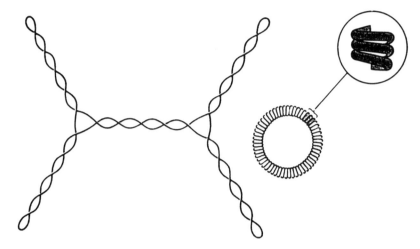

Fig. 4. Negative supercoiling. Shown are interwound (left) and solenoidal (right) forms of supercoiling of duplex DNA (line) molecules of the same length and linking number. These two forms of supercoiling differ geometrically, not topologically. Free supercoiled DNA in solution adopts the interwound form, but supercoiling around proteins is usually solenoidal.

I wish to emphasize that the complex topology of DNA is essential for the life of all organisms. In particular, it is needed for the process known as DNA replication, whereby a replica of the DNA is made and one copy is passed on to each daughter cell. The most direct evidence for the vital role played by DNA topology is provided by the results of attempts to change the topology of DNA inside of cells. The topology of DNA *in vivo* is set by a remarkable group of enzymes called topoisomerases. These enzymes promote the passage of DNA segments through each other—they appear to make DNA incorporeal—until a stable state is achieved. Examples of some of the reactions catalyzed by topoisomerases are shown in Fig. 5.

The topoisomerases are so called because they interconvert topological isomers of DNA, molecules that differ only in a topological property and not in length or base sequence. A number of clinically important anticancer and antibacterial drugs inhibit topoisomerases and rapidly cause cell death. They are lethal to target cells because they stop DNA replication within a fraction of a second. Evidently, variation in normal cellular DNA topology is inconsistent with life.

The importance of DNA topology has also been shown by genetic methods. Mutations that completely eliminate an essential protein cannot be propagated. Conditionally lethal mutations, however, can be propagated. These have lethal consequences under conditions referred to as restrictive, such as high growth temperature, but not under other conditions, called permissive, such as low growth temperature. Organisms carrying these mutations can be grown at low temperature, with the deleterious effect of the mutation expressed by a shift up in growth temperature. Under restrictive conditions for conditionally lethal topoisomerase mutants, DNA topology changes, DNA replication ceases, and the cells die.

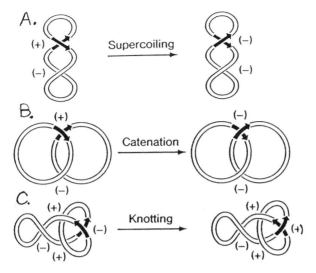

Fig. 5. Reactions of topoisomerases. Duplex DNA is shown as a tube and the arrows indicate orientation. Illustrated are the change in linking number by introducing (–) supercoils (A), forming a (–) singly interlocked catenane (B), and tying a (+) trefoil knot (C). The reverse reactions of removal of supercoils, decatenation, and unknotting are also carried out by topoisomerases. All the reactions are effected by the same operation of passing a segment of duplex DNA through another at the filled arrows. Because these topological changes are brought about by making duplex DNA transiently permeable to the passage of other DNA segments, the enzymes are called type-2 topoisomerases. Type-1 topoisomerases make single-stranded regions of DNA transiently permeable to duplex or single-stranded DNA.

3. How and why DNA topology evolved

Two related questions arise immediately from the recognition that DNA topology is essential for life. How did the complex topology of DNA evolve, and why is it so important for the cell? DNA is the only molecule in cells that has a complex topology. The evolution of proteins has taken a contrary course. Proteins also naturally subdivide into domains and thus local knots or links could readily occur, but they do not. In addition, no knots, catenanes, or supercoiling have been found in RNA, polysaccharides, or lipids.

To answer these evolutionary questions, I shall try to reconstruct history from its end result, the present structure. The answer to the how and why of DNA topology lies in the fundamental (and I believe short-sighted) choice of the one-dimensional genetic code very early during evolution. Essentially all of the genetic information in DNA and RNA is contained in the order of the four different bases. This is a simple way to obtain a reliable, easily duplicated set of blueprints for the cell. The choice of the one-dimensional code was made so long ago that all organisms use basically the same code. Therefore, when the one-dimensional genetic code was first adopted, organisms must have been relatively simple. Increasing complexity of organisms during evolution was necessarily coupled to increases in the length of DNA. The alternatives of evolving a more complex linear code or a 2- or 3-dimensional code were proscribed by the pervasive effects of even a small change in the code. As a result, contemporary organisms have exceedingly long DNA molecules. The DNA in a human cell, for example, is some 10^4 times longer than the diameter of the space it occupies. The challenge for information storage and retrieval is evident.

The well-known double-stranded form of DNA (Fig. 1) is ideally suited to carrying long stretches of the one-dimensional code. Long molecules are easy to break by mechanical, chemical, and enzymatic means. In the case of DNA, it is easy to repair damage caused by breakage of one strand by using the information in the intact strand (Fig. 6). Moreover, most damages to the genetic message are immediately apparent as a distortion in the double helical structure of DNA. A battery of enzymes in the cell scans for such distortions and repairs them.

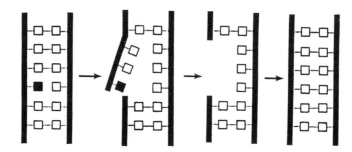

Fig. 6. DNA repair. A short portion of duplex DNA has been untwisted and the backbones are depicted as thick lines and the bases as boxes. The filled box represents the damaged base that will be repaired. The damage disturbs the local structure of the DNA so that it is recognized by an enzyme that breaks the backbone near the damaged base. The region of the DNA containing the damage is removed by another enzyme, leaving a gap in the DNA. The gap is filled with the correct bases and repair is completed. The duplex structure of DNA played three roles in this process. First, the two strands are so close to each other in space that most changes distort the structure enough that they are recognized by repair enzymes. Second, the intact strand maintained continuity while the damaged region was removed. Third, the intact strand provided the information for correct repair. Two physically separate but complementary DNA strands would provide none of these functions.

There are organisms in which single-stranded DNA or RNA is the genetic material. These are exceptions that prove the rule. Only simple viruses have single-stranded genetic information. They rely on their complex host for most of their functions.

The conclusion is that a duplex structure is needed to maintain the fidelity of a long genetic message. But why are the two strands of DNA intertwined helices? The helix is the most common folded form of long polymers in nature. Proteins and RNA, for example, also frequently have helical regions. A helical form for polymers allows the constituent monomers to interact in a simple repetitive fashion that makes an energetically favorable geometry. The requirement for two adjacent helical strands of DNA was solved by intertwining them. The double helical structure also imparts considerable shear strength to DNA and resistance to unwinding and strand separation. It is this intertwining of the two strands of the double helix that is the root cause of DNA topology.

The intertwining of the two strands of DNA directly results in a linking number because DNA is topologically constrained. The value of the linking number per unit length has itself evolved. The linking number for all natural DNA is significantly less than that of the most stable state, the relaxed state. The result is universal (–) DNA supercoiling. As discussed in more detail below, the selective advantage of a deficit in linking number is that it provides energy for the separation of the two strands of the double helix that is necessary for many biological processes.

These two levels of coiling (the double helix and supercoiling), combined with the great length of DNA, lead to the three topological properties of DNA. I will briefly describe four examples of coiling that can be converted to catenation, knotting, or linking number changes. First, the right-handed helical intertwining of the double helix generates catenanes of the same topology at the termination of DNA replication (Fig. 7). Second, knots and catenanes are created by passages of DNA segments through each other by topoisomerases (Fig. 8) and by recombination enzymes (Fig. 9). These knots and catenanes have the same interwound helical form as the substrate supercoils and their complexity is proportional to the number of these supercoils. Third, DNA is compacted by a regular supercoiling around proteins called histones. This geometric change results in a topological change because topoisomerases remove the compensating supercoils outside the bound region. Fourth, the necessity for powerful topoisomerases in the cell leads to accidental knotting and catenation of the tightly packed DNA.

I shall now summarize the evolutionary argument for the origin of DNA topology. Because the genetic code is 1-dimensional, DNA must be very long. Because DNA is so long, it has to be a duplex. The interwound double helix is a compact, stable, energetically favorable form for long duplex DNA. Because the double helix is very long, it gets subdivided into separate topological domains. The direct result is linking of the two strands of the duplex. The indirect result is supercoiling, because the linking number is less than that for relaxed DNA. Coiling and supercoiling, along with the additional coiling needed to pack the DNA into cells, lead to catenation and knotting by enzymes that break and rejoin DNA.

4. Solution of the topological dilemma for the cell

The net selective advantage of DNA topology led to its evolution. However, topological properties have disadvantages that needed to be mitigated during evolution. The most serious of these is that all topological properties of DNA must be removed before the progeny molecules from DNA replication can be segregated into daughter cells. The linking number, catenation, and knotting of DNA must be reduced to

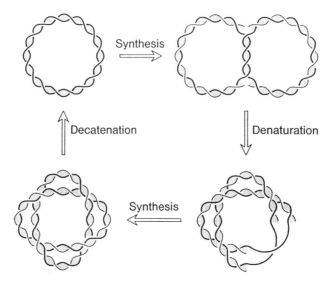

Fig. 7. Conversion of parental DNA helical intertwining into daughter DNA catenation at the terminus of DNA replication. A circular DNA molecule or the topologically constrained domain of a linear molecule is depicted at the upper left. In the first stage of DNA replication (synthesis), a topoisomerase reduces parental DNA linking number concomitant with strand separation and daughter strand synthesis. Eventually steric and topological problems impede this process when the parental strand linking number is of the order of 10. Denaturation (strand separation) of the remaining parental duplex then converts the molecule into a catenane of two gapped DNA rings. After the completion of daughter DNA synthesis, the intact duplex catenanes are unlinked by a type-2 topoisomerase. Replication is now complete.

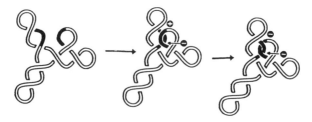

Fig. 8. Conversion of substrate DNA supercoils into knots by a topoisomerase. The topoisomerase is shown acting on a (–) supercoiled DNA (tube). In the first step, two segments of DNA cross each other at the filled regions. In the second step, strand passage at a crossing generates a knot. The number of crossings in the knot is proportional to substrate supercoiling density.

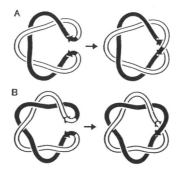

Fig. 9. Metamorphosis of supercoils into knots and catenanes by recombination. The DNA is shown as a tube. Crossover between the sites for recombination (arrows) results in knotting if the sites are oriented along the DNA contour head-to-head (A) and catenation if the sites are oriented head-to-tail (B). The right-handed intertwining of the substrate supercoils is preserved in both products.

on the order of 250 million per human cell. Moreover, populations of DNA molecules do not have a uniform linking number but a Boltzman distribution of values. For nearly relaxed molecules, the distribution is wide because the energy differences between different topoisomers are very small. However, a distribution in linking number about zero will not permit segregation of daughter DNA, because most molecules will have a nonzero value of linking number. Moreover, linking number is a global property, and it is difficult to conceive how an enzyme, which must act locally, can reduce linking number to exactly zero. The problem is further exacerbated by a physical constraint. The topoisomerases that reduce the linking number of the parental DNA act ahead of the growing point on the unreplicated region (Fig. 7). As DNA replication proceeds, the unduplicated region gets progressively smaller until it is too small to be bound and acted on by topoisomerases. Decatenation and unknotting are as great a problem as linking number reduction. The nucleus, the cellular compartment for the genetic material, is choking with DNA. Given such a high DNA concentration and the presence in the nucleus of agents that further condense the DNA, how can random passages by topoisomerases lead to total decatenation and unknotting?

These are profound questions and we have only partial answers. However, what we do know illustrates well the important features of DNA topology in cells. For the reasons already given, any solutions to the topological dilemma cannot compromise the duplex structure of DNA. Therefore, the conditions inside the cell cannot be set by evolution to reduce the linking number of all DNA to zero. Instead linking number reduction is coupled to DNA replication, as illustrated in Fig. 7. Proteins that are part of the replication apparatus continuously unwind the double helix a short distance ahead of the growing point. This creates positive superhelical stress, which is removed by a topoisomerase. Thus, negative superhelicity is maintained at about its prereplication level and linking number is reduced *pari passu* with DNA synthesis.

This scheme efficiently removes most of the linking of DNA. The abovementioned difficulties in reducing the linking number to exactly zero are still present near the terminus of replication. Instead of being removed, the last double helical windings are unwound and converted to catenane links (Fig. 7). The catenanes are then unlinked by topoisomerases and the topological dilemma is solved. Unlike the case for linking number, there is no penalty for setting the cellular conditions such that catenation and knotting are always removed. In fact, it is highly desirable.

5. Forces favoring decatenation and unknotting *in vivo*

Thus, nature solved the linking number dilemma at the termination of replication by converting the linking to a form it could handle, catenation. This introduces a new question. What forces favor decatenation but leave linking number untouched? First, decatenation removes the bending energy needed to wind the two DNA rings around each other. Second, the removal of the last catenane link is entropically favorable because it permits unlinked states. The magnitude of this effect depends on the effective intracellular DNA concentration, and I speculate that evolution has kept this value below the actual physical DNA concentration by keeping DNA hidden behind proteins. Finally, strong mechanical forces exist that pull apart the daughter DNA molecules in chromosomes. Devices called the mitotic and meiotic spindles accomplish this in higher organisms, and perhaps, so does any attachment of daughter DNA to parts of the cell that are growing apart.

The importance of these mechanical forces in decatenation has strong experimental support. In all cells, one topoisomerase has the primary task of unlinking catenanes resulting from DNA replication. Inhibition of the action of this enzyme by drugs or mutations causes cell death and accumulation of catenanes of circular DNA molecules (Fig. 7). The effect of topoisomerase inhibition on long linear DNA is even more dramatic. The mechanical forces keep pulling on the interwound DNA until the DNA breaks and spills out of the cells. Simultaneous inhibition of these mechanical forces by disruption of the spindle blocks the breakage of DNA. These results also provide strong support for the presence of topologically constrained domains in linear chromosomal DNA.

Once mechanisms for efficient decatenation or unknotting evolved, catenation or knotting from any source was no longer a problem, and reactions that produced linked forms could evolve. For example, recombination systems that act on short specific DNA sequences, called site-specific recombination systems, usually produce catenanes or knots (Fig 8). The chief reason for this is that the recombination process converts DNA supercoil crossings into knot or catenane crossings. The unlinking of knots and catenanes by topoisomerases provides an energetic driving force for recombination by selectively removing its products.

A corollary of this line of reasoning is that contemporary site-specific recombination systems must have evolved after topoisomerases. This conclusion is supported by two independent arguments. First, many site-specific recombination systems require negative supercoiling for their activity. Because topoisomerases are needed to generate supercoiling, they had to precede site-specific recombination enzymes. Second, the structure of site specific recombination enzymes strongly suggests that they evolved from the fusion of a gene that encodes a topoisomerase with a gene that encodes a site-specific binding protein. Recombination enzymes have a topoisomerase activity that is made site-specific by the action of a separate domain within the protein that binds to specific DNA sequences.

6. The importance of negative supercoiling

It is easy to understand why topoisomerases have been evolved to unlink catenanes and knots, but they have an equally important role (which probably evolved later) in maintaining (–) supercoiling. The key to understanding supercoiling is to recognize that supercoils have a topological sign opposite to the winding of strands in the double helix. Negative supercoiling provides a tremendous amount of energy for local strand separation. Such unwinding of the double helix results in the reduction of relaxed DNA causes (+) supercoiling and makes the molecule more stressed. Local strand separation is important because most of the information-bearing surfaces of the bases are in the interior of the DNA duplex. As a result, essential cellular processes require single-stranded DNA or RNA.

Let me give examples of the importance of being single-stranded. A critical reaction in the cell is the faithful synthesis of specific proteins. This process is called translation, because the language in which proteins are written (amino acid sequence) is very different from that of nucleic acids (base sequence). The nucleic acid that guides translation is RNA, not DNA. This RNA must be single-stranded during translation because it forms a transient duplex with an adapter molecule that is part nucleic acid and part amino acid. The RNA that guides translation is copied from DNA in a process called transcription because the same genetic message is rewritten in similar but distinct genetic language. Transcription requires the local unwinding of duplex DNA into a single-stranded form from which a complementary RNA is synthesized. This process is greatly favored by negative supercoiling.

Many proteins bind preferentially to single-stranded regions in DNA formed by local unwinding. This process is also favored by (–) supercoiling. In extreme examples, binding of such proteins requires that the DNA be (–) supercoiled. I have called these proteins "supercoil parasites" because they cannot bind to relaxed DNA.

Thus, negative supercoiling is a mechanism that nature has chosen to compensate for the topological penalty for DNA being a double helix. The cell can have its cake (duplex DNA) and eat it (single-stranded DNA during active metabolism).

The density of negative supercoiling *in vivo* is set so that the DNA duplex is just about to unwind locally. A little more supercoiling leads to spontaneous helix unwinding. This would be undesirable because the cell wants to control helix opening and thereby control processes such as transcription and replication. The duplex form of DNA hides the genetic information from enzymatic purview until specific proteins, with the aid of (–) supercoiling, locally unwind the DNA. This important advantage of the duplex structure of DNA has greatly aided the evolution of complexity but, I believe, was not the original source of selection pressure for the double helix.

Supercoils are important in another way. Supercoiled DNA in solution has an interwound conformation in which DNA sites that are far apart along the contour length can be close together in space (Fig. 4, left). Because many protein complexes that act on DNA bind simultaneously to more than one site, supercoiling greatly aids the formation of these complexes. A good example is provided once again by site-specific recombination. Before the breakage and reunion of DNA, two recombination sites must come together on the surface of the recombination enzyme. This juxtaposition is greatly facilitated by the conformation of negatively supercoiled DNA.

Both the energetic and conformational roles of supercoiling require that the supercoiling be maintained only by a topological constraint—a deficiency in linking number, and not by a geometric constraint—winding around proteins. Protein-bound supercoils do not have the conformational freedom needed for site juxtaposition or local strand separation. However, supercoiling formed by winding DNA around proteins is just as important in nature as free supercoils. It has the very different purpose of compacting DNA for storage. In order to pack the enormously long linear DNA into cells, it is coiled successively, about 4 times (Fig. 10). Each order of coiling shortens the DNA about ten times so that the total compaction is about 10^4.

This compaction is readily reversed when the DNA is transcribed, recombined, or replicated.

Supercoiling around proteins provides the first order of coiling, called nucleosome formation (Fig. 10), and perhaps some of the higher orders. The energetic cost of the tight supercoiling in nucleosomes is paid for by the energetic benefit of protein DNA interactions. In topologically constrained DNA, the regular (–) supercoiling in nucleosomes causes compensatory (+) supercoils elsewhere. The energetic cost of supercoiling goes up with about the square of the number of supercoils. Thus, after only a few nucleosomes are formed, the energy of putting one more (+) supercoil into the DNA would equal the protein-DNA interaction energy and nucleosome formation would cease. The topoisomerases come to the rescue and remove the (+) supercoils so that all the DNA can be coiled around nucleosomes.

Nucleosomes are a quiescent storage form of supercoiling. However, when the histones are removed, the DNA explodes into the active free form of (-) supercoiling that we have already discussed. Consequently, an equilibrium exists between the compact storage form of supercoiling and the extended, dynamic, free form.

7. Synopsis of the evolution of DNA topology

I can summarize many of the concepts presented in this article by imagining how the genetic material might have evolved. For simplicity, I will present a single temporal order for evolution, but the process is a network with simultaneous pathways.

Most biologists believe that single-stranded RNA evolved before both proteins and DNA. A major evolutionary step was the direction of protein synthesis by RNA in the translation process. The result was more sophisticated catalysts made of protein and an accelerated pace to evolution. RNA molecules today are at most 100 kb in length. This is but a very small fraction of the coding capacity needed for even a primitive one-celled organism. Greater coding capacity can be achieved by increasing the number of different coding molecules, but this route is limited by the necessity of providing each daughter cell with at least one copy of each coding molecule. Early in evolution, the danger of loss of genetic material was probably mitigated by increasing the numbers of each coding molecule. There is safety in numbers! However, even with the elaborate mechanism for orderly segregation of daughter DNA involving mitotic and meiotic spindles in contemporary organisms, the number of different DNA molecules is only of the order of 10^2. Another problem for the RNA genome may have been that many enzymes have evolved that modify or break down RNA. While this is advantageous for a molecule that is in the thick of metabolism, it presents dangers to the repository of genetic information.

The metabolic activity of RNA also presented an opportunity. A modified form of RNA evolved that had the same 1-dimensional code, but was less reactive. It is called DNA. DNA differs chemically from RNA in only one key respect: it is missing the single, exposed, highly reactive hydroxyl group on each monomer of RNA. This single modification causes a large change in nucleic acid conformation so that enzymes can easily distinguish DNA from RNA. DNA could then be fashioned into a duplex form, leaving RNA to fulfill many of the roles of a single-stranded nucleic acid.

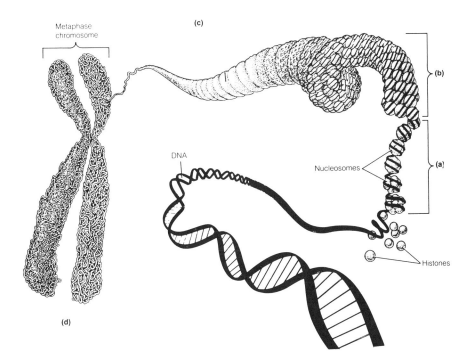

Fig. 10. DNA compaction inside cells by successive orders of coiling. A DNA double helix is compacted in about four successive steps (a–d). Only the first step of nucleosome formation (a) is well understood. In this step, DNA coils twice in a left-handed helical fashion around a set of proteins called histones. The nucleosomes are then coiled successively to give the final form, called a chromosome (d). From Biochemistry by Mathews and van Holde (Redwood City, CA: Benjamin/Cummings Publishing Company, 1990), p. 1006. Reprinted by permission.

With the evolution of type-1 topoisomerases, compaction by nucleosomes could occur and the size of DNA could grow to about 10^5 kb. However, as DNA grew in length, the problems of accidental knotting within domains and accidental catenation of separate domains and segregation of the products of DNA replication became acute. These problems were solved by the evolution of the type-2 topoisomerases, which promote the passage of duplex DNA through transient double strand breaks (Fig. 5). A type-2 topoisomerase could have evolved from a type-1 topoisomerase by the development of an interaction between two copies of a type-1 enzyme. Further increases in DNA size required only the evolution of successively higher orders of DNA compaction.

In order for organisms to become more complex, the DNA had to grow in length. As long as the DNA length was less than the diameter of a nucleus, roughly 200 kb, there was no compaction problem. A simple polymer chemistry calculation based on the bending stiffness of DNA leads to the conclusion that even a 3×10^4 kb

DNA molecule could readily fold into the nucleus. Further evolution demanded longer DNA molecules and, therefore, a first order of compaction. The winding of DNA around nucleosomes compacts DNA molecules about 6 times. Given that the typical size of a topologically closed domain is 100 kb, DNA at this stage of evolution would be folded into many domains. Domains, however, would prevent extensive supercoiling around nucleosomes because of the buildup of compensatory (+) supercoils. This could be circumvented initially in evolution by enzymes that break one of the two DNA strands, called single strand nucleases, followed by sealing the break. However, this is a rather haphazard and hazardous way to relieve the (+) supercoiling stress.

Here, once again, a clumsy early solution to a problem presented an opportunity for the evolution of a more effective one. Enzymes usually break DNA in two steps (Fig. 11A). In the first step the phosphodiester bond that holds DNA together is broken by replacing it with a bond to the enzyme. In the second step, water breaks the DNA-enzyme ester bond. The net result is that water breaks the DNA but the enzyme makes the reaction much faster and more specific. However, water need not do the breaking. The reactivity of water is due to a hydroxyl group. When the enzyme breaks DNA in the first step, it exposes a hydroxyl group in the DNA itself. If, during evolution, this DNA hydroxyl which is released in the first step of the reaction becomes the attacking group in the second step, then the net result is a transient single-strand break in DNA (Fig. 11B). Since supercoiling is a strained state, relaxation could readily occur between the nicking and sealing steps. I have just described a mechanism of type-1 topoisomerases. It is easy then to see how type-1 topoisomerases might have evolved from nucleases. The topoisomerases perform the magician's trick of breaking the DNA but holding on to both broken ends so that the danger of interruption of the genetic material is minimized.

8.   Catenated DNA networks in kinetoplasts

I shall finish this article with an exception to this global picture that serves, nonetheless, to illustrate general rules. There is a family of clinically important unicellular parasites that has an intracellular compartment called a kinetoplast that contains a uniquely structured DNA. The DNA in the kinetoplast is a catenated network of 10^4 DNA rings and violates almost every rule I have stated. This catenane is very stable, and the constituent DNA rings are neither supercoiled nor compacted by incorporation into nucleosomes. I believe that this kinetoplast DNA represents a primitive alternative to the evolution of long DNA molecules. The catenation of these DNA rings prevents them from getting lost during cell division, just as a long single DNA molecule keeps a multitude of genes together. The 2-dimensional kinetoplast DNA network (Fig. 12) actually brings about much more efficient compaction of DNA than the 1-dimensional winding of DNA around nucleosomes. Supercoiling, in fact, inhibits network formation, because the smaller cross-sectional area of supercoiled DNA makes it harder to catenate every ring to three other rings, as found in kinetoplast DNA (Fig. 12). The kinetoplast network is rather primitive in that it contains several different rings, each in multiple copies that seem to be randomly located throughout the network. It could have evolved to an advanced genetic material if the 1-, 2- and 3-dimensional ordering of the rings had an information content. Maybe in another galaxy, it does.

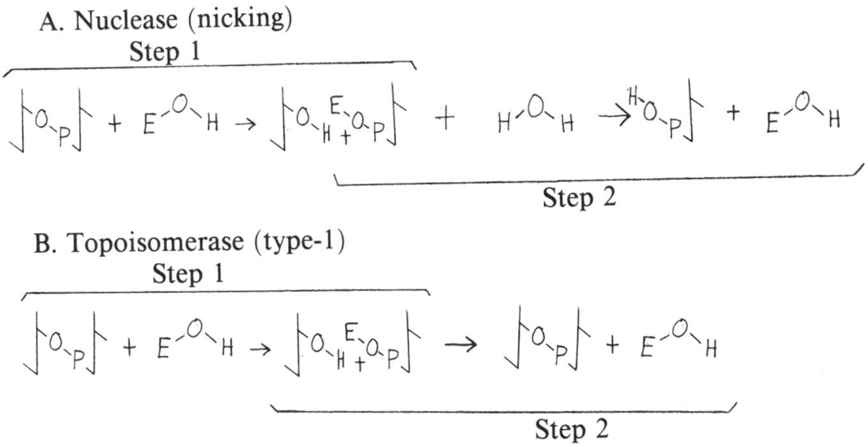

A. Nuclease (nicking)

Step 1

B. Topoisomerase (type-1)

Step 1

Fig. 11. Comparison of a nuclease with a topoisomerase. Only one backbone bond, called a phosphodiester bond, of a long duplex DNA chain is shown.   A. Nuclease activity.   In step 1, a hydroxyl group on an enzyme, E, attacks the phosphodiester bond, interrupting one strand of the double helix and forming an enzyme-DNA phosphoester. In step 2, a hydroxyl group of water attacks this bond to regenerate the enzyme and complete the break in DNA.   B. Topoisomerase activity. Step 1 is the same as in A, but in step 2 the neighboring DNA hydroxyl attacks the enzyme-DNA phosphoester to regenerate both the enzyme and intact DNA.   If between steps 1 and 2, a DNA strand passed through the transient break, then the topology of the DNA is changed.

Fig. 12.   A section of a kinetoplast DNA network.   A small section on the edge of a kinetoplast DNA network is shown.   Except for the DNA rings at the edge, each ring is catenated via a single interlock to three other rings.   The result is an ordered sheet of DNA rings.

REFERENCES

The following is a short annotated list of reviews and expository articles that provide good introductions to the topics considered:

1. Benner, S. A., Ellington, A.D., and Tauer, A.   Modern metabolism as a palimpsest of the RNA world. Proc. Nat'l. Acad. Sci. USA **86** (1989), 7054-7058. A consideration of life before the ascendancy of DNA as the genetic material. The concept of the RNA world is among the most important contributions of molecular biology to evolutionary theory.

2. Cozzarelli, N. R., Boles, T.C., and White, J.H.   A primer on the topology and geometry of DNA supercoiling . In: N.R. Cozzarelli and J.C. Wang, eds., *DNA Topology and Its Biological Effects.* Cold Spring Harbor Laboratory Press, Cold Spring Harbor, N.Y., 1990, pp. 139-184. A comprehensive treatment of the forms and quantitative descriptors of DNA supercoiling.

3. Kornberg, A. and Baker, T.A.   *DNA Replication* , W. H. Freeman and Company, New York, 1991.   An up-to-date textbook.   Chapter 12 is a good introduction to topoisomerases.

4. Wang, J.C. (1987).   DNA topoisomerases: Nature's solution to the topological ramifications of the double-helix structure of DNA. The Harvey Lectures, **81**, 93-110. Easy-to-read short essay on the teleology of topoisomerases.

5. Wang, J.C. and Liu, L.F.   DNA replication: topological aspects and the roles of DNA topoisomerases. In: N.R. Cozzarelli and J.C. Wang, eds., *DNA Topology and Its Biological Effects.* Cold Spring Harbor Laboratory Press, Cold Spring Harbor, N.Y., 1990, pp. 321-340. Up-to-date and very clear.

6. Wasserman, S.A. and Cozzarelli, N.R. Biochemical topology: Applications to DNA recombination and replication.   Science **232** (1986), 951-960.   Very good short introduction to the role of topology in biology.

7. White, J.H.   An introduction to the geometry and topology of DNA structure. In: M. S. Waterman, ed., *Mathematical Methods for DNA Sequences,* CRC Press, Boca Raton, FL, 1989, pp. 225-253. An authoritative and rigorous treatment by a mathematician.

DEPARTMENT OF MOLECULAR AND CELL BIOLOGY, DIVISION OF BIOCHEMISTRY AND MOLECULAR BIOLOGY, UNIVERSITY OF CALIFORNIA, BERKELEY, CALIFORNIA 94720.

E-mail address: cozzlab@garnet.berkeley.edu

Proceedings of Symposia in Applied Mathematics
Volume **45**, 1992

Geometry and Topology of DNA and DNA - protein interactions

James H. White

ABSTRACT. The application of geometric and topological methods to molecular biology has provided a new understanding of the relation between the twisted paired strands of the genetic molecule DNA and its overall supercoiled structure. In addition, the geometric and topological changes in the DNA caused by the interaction of protein complexes such as nucleosomes with the DNA can now be quantized. The linking number, writhing number, and twist are defined and used to describe the geometric and topological properties of closed supercoiled DNA. In addition, it is shown by geometric techniques that the linking number of a closed circular DNA associated with a protein complex is the sum of two experimentally accessible quantities: the winding number determined by the protein complex, a number directly related to the helical repeat, and the surface linking number, a geometric constant that accounts for the topological effects of the protein complex on the super-coiling of the DNA. These results are applied to experiments involving x-ray diffraction, nuclease digestion and chemical probes of DNA wrapped on nucleosomes.

1. DNA Geometry and Topology

DNA is usually envisioned as a pair of cylindrical helices, **C** and **W**, representing the sugar-phosphate backbones, winding about a straight linear axis, **A**, (Figure 1A). However, it is now clear that many DNA are closed molecules, i.e., their axes as well as their backbone curves are closed curves, (Figure 1B). In fact, the axes of these closed DNA can assume almost any path in space. It has been found by direct experiment that most closed DNA have the shape shown in Figure 1C. Such DNA are called *supercoiled* since the axis is seen to coil back on itself and so the backbone helices are supercoiled [1]. This lecture is concerned with the geometry and topology of closed supercoiled DNA. We first define mathematical quantities which can be used to describe the physical shape of the supercoiling as well as experimental results.

1991 Mathematics Subject Classification. 53A05, 57D2X, 92XX
Supported by NSF Grant DMS-8720208
This paper is in preliminary form and will be published elsewhere.

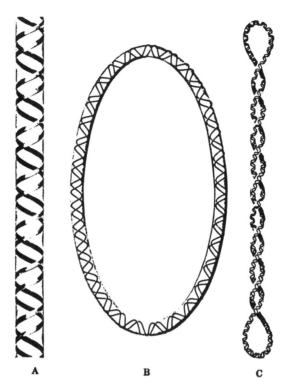

Figure 1. A. The linear form of the double helical model of DNA. B. The relaxed closed circular form of DNA. C. The plectonemically interwound form of supercoiled closed DNA

       Three quantities most useful for describing closed DNA free in solution are *linking, writhe,* and *twist,* [11] The linking number is a mathematical quantity associated with two closed oriented curves. To define, it the simplest manner is to use the so-called modified projection method. We designate the two curves **C** and **A**. These two curves when viewed from a distant point will appear to be projected into a plane perpendicular to the line of sight, except that the relative overlay of crossing segments is clearly observable. Such a view gives a modified projection of the pair of curves. In any such projection, there may be a number of crossings. To each such crossing is attached a number ± 1, depending on the sign convention in Figure 2. Adding all the signed numbers of a given projection and dividing by 2 gives the linking number, Lk(**C**,**A**) of **C** with **A**. Examples are shown in Figure 3.

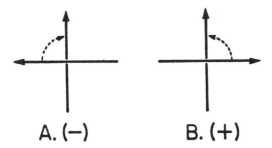

Figure 2. Sign convention for the crossing of two curves in a modified projection. The arrows indicate the orientation of the two crossing curves. To determine the sign of the crossing, the arrow on top is rotated by an angle less than 180° onto the arrow on the bottom. If the rotation required is clockwise as in A, the crossing is given a (-) sign. If the rotation required is counterclockwise as in B, the crossing is given a (+) sign.

The linking number has many important properties, two of which are especially important for DNA. First, it is unchanged under any continuous deformation of the pair of curves so long as no break is made in either curve. Second, it is independent of the view for which one computes it. For DNA the linking number is defined to be the linking number of the two backbone curves. However, since either backbone curve may be deformed into the axis curve **A** without passing through the other, the linking number of a DNA may equally be defined as the linking number of a backbone curve and the axis, Lk(**C**,**A**). This latter definition is the one which we will use in this lecture.

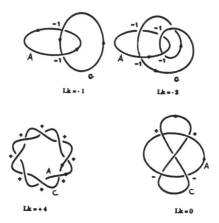

Figure 3. Examples of pairs of curves with various linking numbers, using the convention described in Figure 2 and the method described in the text.

The linking number of a DNA, though a topological quantity, divides into two geometric quantities, writhe, Wr, and twist, Tw which are useful in describing the geometric properties of supercoiling [10]. As we have seen, the linking number is a measure of the number of crossings of the pair of curves, **C** and **A**, in any view. These crossings may be subdivided into distant crossings and local crossings. Distant crossings occur when the axis is seen to cross itself, for in this case the backbone curve **C** of one portion of the DNA crosses the axis curve **A** of another distant portion. Local crossings occur from the winding of the backbone curve **C** about the axis curve **A** in the same portion of the DNA. Writhe is a measure of the distant crossings and twist is measure of the local crossings. We now give precise definitions of these two quantities.

The definition of writhe is similar to that of linking. However, it is a property of a single curve, in this case the axis **A**. In any modified projection of **A** there may be a number of crossings. To each such crossing is attached a signed number ± 1 as in the case of linking number. If one adds all these numbers one obtains the projected writhing number. Unlike linking number, the projected writhing number may depend on the projection. This is illustrated in Figure 4 in which a figure eight in one projection becomes an oval curve in another. In one case the projected writhe is -1, in the other 0. The writhing number or writhe, Wr, of the curve **A** is defined to be the average over all possible projections of the projected writhing number. In our example the writhing number is therefore somewhere between -1 and 0.

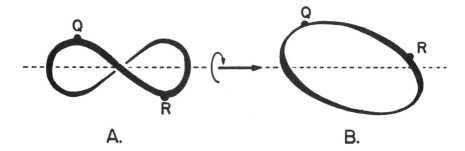

A.   B.

Figure 4. Illustration of the dependence of projected writhing number on projection. The axis of the same non-planar closed DNA is shown in two different projections obtained by rotating the molecule about the dashed line. The points Q and R on the axis help illustrate the rotation. The segment QR crosses in front in part A but is in the upper rear in part B. The projected writhing number in part A is -1 and 0 in part B.

If two different segments of the axis curve **A** are brought very close together, then the proximity results in a crossing which is seen in almost all modified projections. Thus this proximity will result in a contribution to Wr of approximately ± 1. If the axis lies in a plane, then Wr = 0. This is due to the fact that in all projections (except, of course, along the plane itself) there will be no crossings seen. If **A** lies in a plane except for a few places at which it crosses itself, Wr is the total of the signed numbers attached to the self-crossings. Figure 5 illustrates the approximate writhe of some tightly coiled DNA axes. An important fact about the writhing number is that during a self-passage of the curve **A** it must change by 2. This is illustrated in Figure 6. We also point out that the writhing number of a curve **A** is independent of the orientation chosen along the curve.

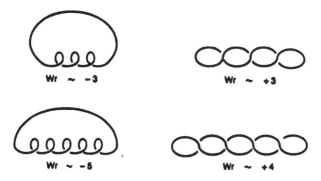

Figure 5. Examples of closed curves with different writhing numbers.

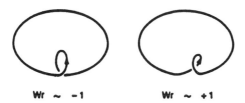

Figure 6. The writhing number of curves with one coil. The curve on the left has writhing number approximately -1 and on the right approximately +1. One curve may be obtained from the other by a self-passage at the crossing changing the writhing number by +2 or -2.

We next define the twist of a DNA. For closed DNA the twist will usually refer to the twist of one of the backbone curves **C** about the axis curve **A**. This will be denoted Tw(**C**,**A**) or simply Tw. To define the twist we use vector analysis [12].

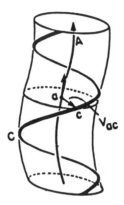

Figure 7. Cross-section of a DNA. The plane perpendicular to the DNA axis **A** intersects the axis in the point a and the backbone curve **C** in the point c. The unit vector along the line joining a to c is denoted $\mathbf{v_{ac}}$. Note that as the intersection plane moves along the DNA, this vector turns about the axis.

Any local cross-section of a DNA perpendicular to the axis **A** contains a unique point a of the axis and a unique point c of the backbone curve **C**. (Figure 7) We denote by $\mathbf{v_{ac}}$ a unit vector along the line joining a to c. As the DNA is traversed since the curve **C** winds helically about **A**, the vector $\mathbf{v_{ac}}$ turns about **A**. Tw is a measure of this turning. As the point a moves along **A**, the vector $\mathbf{v_{ac}}$ changes. The infinitesimal change in $\mathbf{v_{ac}}$, denoted $d\mathbf{v_{ac}}$, will have a component tangent to the axis and a component perpendicular to the axis. Tw is the measure of the total perpendicular component of the change of $\mathbf{v_{ac}}$ as the point a traverses the entire length of the DNA. This is given by the line integral:

$$\text{Tw} = \frac{1}{2\pi}\int_A d\mathbf{v_{ac}} \cdot \mathbf{T} \times \mathbf{v_{ac}},$$

where T is the unit tangent vector along the curve **A**. When **A** is a straight line or planar, $d\mathbf{v_{ac}}$ is always perpendicular to **A**, so that in this case Tw is simply the number of times that $\mathbf{v_{ac}}$ winds about the axis. Examples are shown in Figure 8. It can be easily demonstrated that Tw is positive if the winding is right-handed and negative if left-handed. Furthermore, if the DNA is closed then the initial and final positions of $\mathbf{v_{ac}}$ are the same. Thus if the DNA is closed and its axis planar, Tw must necessarily be an integer. However, if the axis is supercoiled this is not usually the case [12]. A portion of a supercoiled DNA is shown in Figure 8. Here the axis **A** itself is a helix, so that the helically winding **C** becomes a superhelix. For such an example, Tw is the number of times that **C** winds about **A** plus a term $n\sin\gamma$ which depends on the geometry of the helix **A**. n is the number of times that **A** winds about its owns straight line axis and γ is its pitch angle.

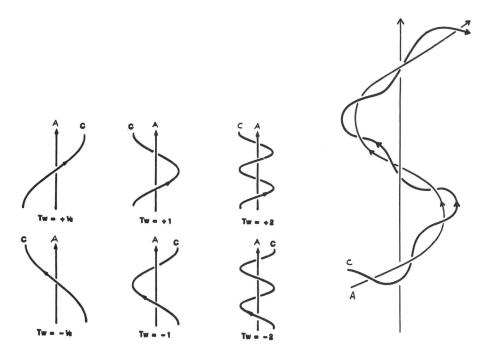

Figure 8. Examples of pairs of curves **C** and **A** with different values of twist. The first six are simple examples in which the axis **A** is a straight line and the twist is the number of times that **C** winds about **A**, being positive for right-handed twist and negative for left-handed twist. The last example is one in which the axis **A** is a helix winding around a linear axis, and the curve **C** is a superhelix winding about **A**. In this case the twist of **C** about **A** is the number of times that **C** winds about **A** (in this case approxiamtely 3.5) *plus* $n\sin\gamma$, where n is the number of times that **A** winds about the linear axis and γ is the pitch angle of the helix **A**. Here n is approximately 1.5 and γ is approximately 40° Thus, Tw equals approximately 4.46.

As we have seen in any modified projection of a pair of curves **C** and **A**, the crossings which are counted to give the linking number divide into distant crossings and local crossings. Projected writhe arises from distant crossings and in the simple cases shown in Figure 8 each complete twist revolution of **C** about **A** gives rise to two local crossings of the same sign. This intuitive idea may be formally expressed by the following theorem [10]:

$$Lk = Tw + Wr.$$

This theorem is a special case of Theorem 1 in the appendix in which **v** is chosen to be along v_{ac}. The remarkable fact about this result is that two geometric quantities that may change under deformations of the curves sum to a topological quantity which is invariant under such deformations. (It is important to remember that the projected writhe may change depending on the view.) Thus for a closed DNA of constant linking number, any change in Wr that occurs during a deformation must be compensated by an equal in magnitude but opposite in sign change in Tw. This interchange is most easily seen by taking a rubber band or some elastic like ribbon material with two edges and while holding one segment fixed with one hand, begin twisting another with the other hand. After a time the rubber band will be seen to coil upon itself, indicating the onset of writhing. The linking number of the edges is of course unchanged since the rubber band is not broken during the twisting. Hence, the twisting must be compensated by writhing. Though more complicated to explain, it is the same phenomenon that accounts for the supercoiling of most heavily used telephone cords. The constant twisting of the cord is eventually compensated by writhing. Taking the receiver off the hook and letting it hang freely usually results in it rotating rapidly to change the writhing by untwisting.

A simple application of this theorem to DNA is an explanation of its propensity to supercoil. Figure 1B depicts a closed circular DNA in its so-called relaxed state. In this case the axis **A** is planar so that its writhing number, Wr, is equal to zero. Therefore, by the fundamental formula, Tw = Lk. Thus, both Tw and Lk are equal to the number of times that either the backbone strands winds about the axis. For such a closed relaxed DNA in the B-form the number of base pairs per turn of the helical backbone is about 10.5. This linking number is usually denoted Lk_0. An example is the simian virus SV40 molecule which has approximately 5250 base pairs. In its relaxed state, $Lk_0 = 500$. However the linking number, Lk, of most closed DNA is not that of the relaxed state. The actual linking number is usually less than Lk_0. In an electron microscope such DNA appear to be contorted or coiled up rings with many self-crossings. This is because the equilibrium state of such molecules in solution have the form shown in Figure 1C. The quantity, $Lk - Lk_0 = \Delta Lk$ is called *the linking number difference* and is a measure of supercoiling. Any change of linking from the relaxed state must divide into a change in twist and in writhing number, $\Delta Lk = \Delta Tw$

+ ΔWr. Since the writhing number of a planar DNA is 0, ΔWr becomes simply Wr. Recent work [2] shows the ratio of the change in writhe to the change in twist is approximately 2.6 to 1, i.e., ΔWr = 0.72(ΔLk) and ΔTw = 0.28(ΔLk). Thus, for each change of 1 in the linking difference, there is a change of 0.72 in the writhe. Large changes in linking will therefore result in large changes in Wr. Because of this, there are negative cross-overs introduced in many views of the DNA. In fact, it has been shown using energetics that the interwound coils shown in Figure 1C well model negatively supercoiled DNA. Such DNA are also called *underwound* because the twist is also reduced. In its native free state SV40 has ΔLk = -25, so that Wr is approximately -18. Thus in some views one will see at least 18 negative crossings.

2. DNA on Protein Complexes

We now turn our attention to the geometric and topological analysis of DNA whose axes are constrained to lie on surfaces [14]. The best characterized example of a protein surface is the nucleosome core [5], a cylinder of height 5.04 nm and radius 4.3 nm. In this case the axis A of the DNA wraps nearly twice around the core as a left handed helix of pitch 2.8 nm. In general the surface of a protein is defined as follows. The surface on which the DNA molecule lies is the so-called solvent accessible surface [6]. (This is the surface generated by moving a water-sized spherical probe around the atomic surface of the protein at the Van der Waals distance of all external atoms, and is the continuous sheet defined by the locus of the center of the probe.) It is this surface that comes in to contact with the DNA backbone chain. Since the DNA is approximately 1nm in radius, the DNA axis lies on a surface that is 1nm outside of the solvent accessible surface to account for the separation of the backbone from the axis. This latter surface is the one to which we shall refer in the rest of this section as the surface on which the DNA, meaning the DNA axis A, lies or wraps.

For DNA that lies on a surface, the geometric and topological analyses are best served by dividing the linking number not into twist and writhe which relate to only spatial properties of the DNA, but into components that relate directly to the surface and surface-related experiments. The linking number of a closed DNA constrained to lie on the surface divides into two integral quantities, the *surface linking number*, which measures the wrapping of the DNA about the surface and the *winding number*, which is a measure of the number of times that the backbone contacts or rises away from the surface [14]. Experimentally the first quantity may be measured by x-ray diffraction and the second may be measured by digestion or footprinting methods. In particular for the nucleosome, the partial contribution to the surface linking number due to the left-handed wrapping about the cylindrical core is -1.85 [4][7]. Furthermore, the winding number has been measured to be the number of base pairs of the DNA on the nucleosome divided by approximately 10.17 [3].

3. The Surface Linking Number

We now give a formal definition of the two quantities, surface linking number and winding number, for a closed DNA on a protein surface. We assume that the surface involved has the property that at each point near the axis of the DNA, there is a well-defined surface normal vector. The unit vector along this vector will be denoted $\mathbf{v}$. (We assume that the surfaces are orientable. In this case there are two possible choices for the vector field $\mathbf{v}$, depending to which side of the surface the vector field points.) If the DNA axis $\mathbf{A}$ is displaced a small distance $\varepsilon \neq 0$ along this vector field at each point, a new curve $\mathbf{A}_\varepsilon$ is created. ε should be chosen small enough so that during the displacement of $\mathbf{A}$ to $\mathbf{A}_\varepsilon$ no crossings of one curve with the other takes place. The curve $\mathbf{A}_\varepsilon$ is also closed and can be oriented in a manner consistent with the orientation of $\mathbf{A}$. The surface linking number is defined to be the linking number of the original curve $\mathbf{A}$ with the curve $\mathbf{A}_\varepsilon$ [13]. Simple examples of the surface linking number occur for DNA whose axes lie on planar surfaces or spheroidal surfaces. First, for a DNA whose axis lies in a plane, SLk = 0. This is easy to see, for in this case the vector field $\mathbf{v}$ is a constant field perpendicular to the plane. Hence the curve $\mathbf{A}_\varepsilon$ lies entirely to one side of the plane and cannot link $\mathbf{A}$, a curve lying entirely in the plane. Second, if the DNA axis lies on a round sphere, SLk = 0. To see this, we can assume without loss of generality that the vector field $\mathbf{v}$ is inward pointing into the sphere. In this case the displaced curve $\mathbf{A}_\varepsilon$ lies entirely inside the sphere and hence cannot link $\mathbf{A}$. These and additional examples are illustrated in Figure 9.

SLk is what is technically called a differential topological invariant and thus has three very important properties. First, if the DNA axis-surface combined structure is deformed in such a way that no discontinuities in the vector field $\mathbf{v}$ occur in the neighborhood of the DNA axis $\mathbf{A}$, and $\mathbf{A}$ itself is not broken, then SLk remains invariant. For example, if the DNA lies in a plane and that plane is deformed, SLk remains equal to 0; or if it lies on a sphere that is deformed, SLk remains equal to 0. Thus, if a DNA axis lies on the surface of any type of spheroid, SLk = 0. Examples of spheroids on shown in Figure 10A. An important example of this is shown in Figure 9D in which a DNA axis is shown to lie on the surface of a capped cylinder. A second important property of SLk is that it only depends on the surface near the axis. Hence, if the surface on which it lies is broken or torn apart away from the axis, SLk still remains invariant. For example, if a DNA lies on a protein, and a portion of the protein away from it is broken or decomposes, SLk remains invariant. The third important property is that if a DNA lies on a surface and slides along the surface, then as long

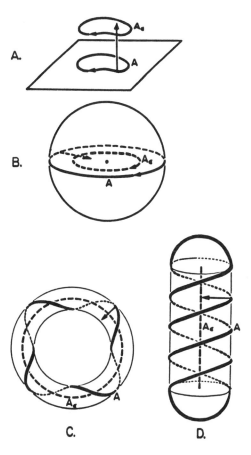

Figure 9. Examples of displacement curves and SLk. For any curve A lying on a surface, the displacement curve A_ε is formed by moving a small distance ε along the surface normal at each point on the curve. For planar curves as in part A, all the normal vectors can be chosen to point upward, and then A_ε is above A. The curves are unlinked and hence SLk = 0. For curves on spherical surface as in part B, the surface vectors can be chosen to point inward, and hence A_ε is entirely inside and therefore does not link A. . SLk is again equal to 0. In parts C and D the surface normal vectors have been chosen to point inward, ε has been set equal to the inner radii of the surfaces on which the DNA is wound. In C, A_ε becomes the central axis of the torus and SLk = +4. In D, the DNA is wrapped plectonemically about a capped cylinder. The displacement curve A_ε lies entirely inside and thus SLk = 0.

as the vector field **v** varies smoothly from point to point in the surface and as long as in the process of sliding, the axis does not break, SLk remains invariant. Thus, if the capped cylinder in Figure 9D were allowed to expand and the axis curve required to remain the same length, it would have to unwind as it slid along the surface of the enlarged cylinder. However, SLk would remain equal to 0.

Another class of biologically important surfaces exist for which it is possible that a DNA can have an SLk $\neq$ 0. These are so-called toroidal surfaces. They consist of the round circular torus and their deformations. Suppose an axis curve A traverses the entire length of a round circular torus handle once as it wraps around it a number of times, n. Suppose further that the inner radius of the torus is equal to r. For the vector field **v**, we choose the inward pointing surface normal. In this case, if one chooses $\varepsilon = r$, A_ε would be the central axis of the torus. Thus SLk is the linking number of the curve **A** with the central axis of the torus. This implies that if the the wrapping is right-handed, SLk = +n and if the wrapping is left-handed SLk = -n. By the invariant properties mentioned above, SLk remains invariant even if the round torus is deformed. Examples are shown in Figure 10B.

The concept of SLk can also be applied to DNA which are not attached to real protein surfaces but which are free in space. For example, the most common kind of free DNA, i.e. DNA free of any protein attachment, is the plectonemically wound DNA. Here the DNA can be considered to lie on the surface of a spheroid such as the one shown in Figure 9D (or a deformation of it), the exact shape of which is determined by the energy minimum DNA conformation. Then the surface may be allowed to vanish and reappear without changing the shape of the DNA superhelix. The DNA is said to be wrapped on a *virtual* surface [14]. Thus, the SLk of the DNA in Figure 9D is equal to 0 regardless of whether the surface is virtual or a that of a real protein. More generally, these concepts may be applied to DNA wrapped on a series of proteins with virtual surfaces joining them. An example of this will be presented later in our discussion of the minichromosome.

4. The Winding Number and Helical Repeat

We next give a formal definition of the winding number of a DNA wrapping on a surface. Since the vector **v** is perpendicular to the surface it is also perpendicular to the DNA axis **A** and thus lies in the

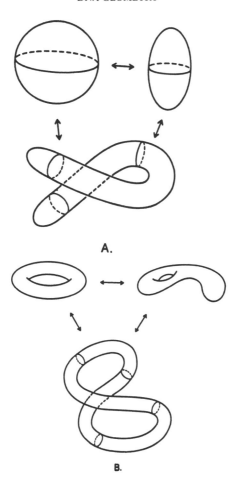

Figure 10. A. Deformations of the round sphere into spheroids. B.
Deformation of the round circular torus into toroids.

perpendicular planar cross-section at each point of the DNA. Therefore,
at each point this vector **v** and the strand-axis vector $\mathbf{v_{ac}}$ defined above
lie in this same planar cross-section.   In this plane, the vector $\mathbf{v_{ac}}$
makes an angle ϕ with the vector **v** (Figure 11). As one proceeds along
the DNA segment $\mathbf{v_{ac}}$ spins about **v**, as the backbone curve **C** alternately
rises away and falls near to the surface, while the angle ϕ turns through
360^{0}.   The total change in the angle ϕ, divided by the normalizing
factor 2π (360^{0} in radians), during this passage is called the winding
number of the DNA and is denoted Φ [14].   This number may also be

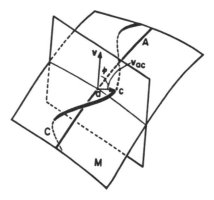

Figure 11. Definition of the surface vectors necessary to define the winding number. The duplex DNA axis **A** lies on the surface. The backbone curve **C** will pass above and below the surface as it winds about **A**. To describe surface winding, two vectors need to be defined originating at a point a on **A**, the unit surface normal vector **v**, and the strand-axis vector $\mathbf{v_{ac}}$ along the line connecting a to its corresponding point c on the backbone **C**. ϕ is defined to be the angle between these two vectors. The winding number Φ measures the number of times that ϕ turns through 360^0, or how many timers $\mathbf{v_{ac}}$ rotates past **v**.

thought of as the number of times that $\mathbf{v_{ac}}$ rotates past **v** as the DNA is traversed. A related quantity called the helical repeat, denoted h, is the number of base pairs necessary for one complete 360^0 revolution. For closed DNA since the beginning and ending point are the same for one complete passage of the DNA, the vectors **v** and $\mathbf{v_{ac}}$ are the same at the beginning and the end. In this case, therefore, Φ must necessarily be an integer.

There are equivalent formulations for the winding number of a closed DNA wrapping on a protein surface. During each 360^0 rotation of the vector $\mathbf{v_{ac}}$ in the perpendicular plane, ϕ assumes the values of 0 and 180^0 exactly once. When $\phi = 0$, $\mathbf{v_{ac}} = \mathbf{v}$, and when $\phi = 180^0$, $\mathbf{v_{ac}} = -\mathbf{v}$. In the former case, the backbone strand comes into contact with the protein, in the latter it is at maximal distance from the protein surface. Thus the winding number of a closed DNA on a protein surface is the number of times one of the backbone strands contacts the protein surface or it is number of times the strand is at maximal distance from it. In this case it also follows from its definition that the helical repeat is the number of base pairs between successive contact points of one of the backbones or the number of basepairs between successive points of maximal distance from the protein surface. This latter number can be measured directly by digestion or footprinting methods which involve probes that search for points of the backbones to cut, the easiest being those points at maximal distance from the surface.

Φ is also a differential topological invariant and therefore has the same three properties mentioned above for SLk. In particular it remains invariant if the DNA-surface structure is deformed without any breaks in the DNA or any introduction of discontinuities in the vector field **v**. Under the same conditions, it also remains unchanged if the DNA is allowed to slide along the surface.

5. Relationship between linking, surface linking, and winding

The three quantities Lk, SLk, and Φ, although very different in definition are remarkably related by a theorem from differential topology. In fact, for a closed DNA on a surface, the linking number is the sum of the surface linking number and the winding number [14], i.e.

$$Lk = SLk + \Phi.$$

This theorem is a special case of Theorem 2 in the Appendix in which case the vector $\mathbf{v_{ac}}$ is to be $\mathbf{v_1}$ and the surface normal vector **v** is chosen to be $\mathbf{v_2}$. For DNA that lies in a plane or on a spheroid, SLk = 0. Therefore, Lk = Φ and if there are N base pairs in the DNA the helical repeat is given by h = N/Lk. These two cases include relaxed circular DNA, for which Lk = Lk_0, and plectonemically interwound DNA, the most common form of supercoiled DNA. For DNA that traverses the handle of a round circular torus while wrapping n times about the handle, Lk = n + Φ if the wrapping is right-handed, and Lk = -n + Φ if the wrapping is left-handed. In both cases Lk is unchanged if the torus if smoothly deformed.

We now outline the proof of the main result. To do this we first define the surface twist, STw, of the vector field **v** along the axis curve **A** [13],[14]. This is defined similarly to the twist of the DNA except that the vector field **v** is used and not the vector field $\mathbf{v_{ac}}$ (refer to the general definition of twist in the Appendix). Hence, STw is given by the equation:

$$STw = \frac{1}{2\pi} \int_A d\mathbf{v} \cdot \mathbf{T_a} \times \mathbf{v}.$$

Thus, STw measures the perpendicular component of the change of the vector **v** as one proceeds along the axis **A**, and thus is a measure of the spinning of the vector field **v** about the curve **A**. It also can be considered to be the twist of $\mathbf{A_\xi}$ about **A**. We recall that Tw measures the spinning of the vector field $\mathbf{v_{ac}}$ about the curve **A**. Thus, the difference Tw - STw measures the spinning of $\mathbf{v_{ac}}$ about **v** (refer to theorem 2 in the Appendix). But this is exactly the winding number Φ. Hence Tw - STw = Φ.

We recall the fundamental formula that:

$$Wr(A) + Tw(C,A) = Lk(C,A) \text{ or}$$

$$Wr + Tw = Lk .$$

A similar formula holds relating STw, Wr, and SLk:

$$Wr(A) + STw = Lk(A,A_\varepsilon), \text{ or}$$

$$= Wr + STw = SLk,$$

since STw is the twist of A_ε about A and SLk is the linking number of A_ε and A. Combining the two formulas and using the result that Tw - STw = Φ, we obtain:

$$Lk = SLk + \Phi.$$

The biological importance of this relationship is that all three of these quantities are experimentally measurable. Thus determining any two of them, one may calculate the other and then compare with the experimental value. We show by a classical example from molecular biology, the minichromosome, the power of this theorem.

6. Application to the Study of the Minichromosome

A minichromosome is a structure which consists of a closed DNA bound to a series of core nucleosomes. Such a structure allows the compactification of a very long DNA into a small volume, in the same way that a long piece of thread is compactified by wrapping it on a spool. Understanding such structures is essential to a knowledge of how DNA is packaged in the cell. In this section we study the geometry and topology of DNA in such a structure. Each nucleosome may best be described as a cylinder, the histone octamer, around which the DNA wraps approximately 1.8 times in a left-handed manner. The DNA segments between successive nucleosomes are called linker regions.

Thus, the DNA divides between linker DNA and core-associated DNA. An example of such a structure is shown in Figure 12 . Such a compound structure consist of a toroidal surface, part of which are the real surfaces of the nucleosomes cores and part of which are virtual linker surfaces joining successive cylinders. These virtual pieces are deformed cylindrical sections, all of the same radius, on which the linker DNA are constrained to lie. The specification of each of these surfaces is arbitrary as long as it takes into account the coiling of the linker. The linker DNA can thus be thought of as a generating curve for the cylindrical section. An important condition to be imposed is that the linker DNA does not wind about the piece on which it lies. This condition will ensure that all contributions to SLk due to winding about the torus handle will come only from the intranucleosome winding. Any additional contribution to SLk must therefore come form the coiling of the linker DNA.

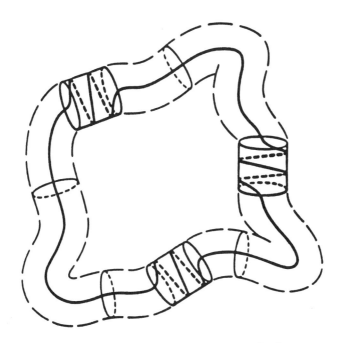

Figure 12. Cartoon of a minichromosome. Three cylinders representing histone octamers are wound by DNA so as to form three nucleosomes. The nucleosomes are connected by linker DNA segments. Successive real nucleosomes are connected by virtual deformed cylindrical pieces, the deformations of which are determined by the coiling of the linker.

To simplify our example we will assume that the minichromosome is relaxed. This means that the linker regions are planar and that all contributions to SLk comes from the winding of the DNA about the histone octamers. Such a relaxed state may be achieved by the introduction of topoisomerases into the minichromosome which relax the linker DNA but leave unaffected the DNA on the nucleosome cores. In this case, then SLk can be directly measured by x-ray diffraction and found to be -1.8m where m is the number of nucleosomes. An example with 5 nucleosomes is shown in Figure 13, for which SLk = -9. For the DNA, SV 40, there are about 25 nucleosomes [9]. Therefore, SLk = -45.

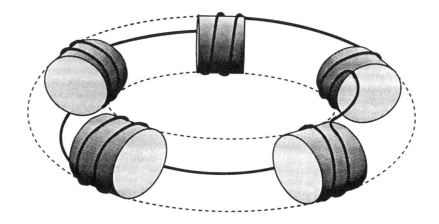

Figure 13. Diagram of a relaxed minichromosome with five cylindrical nucleosomes. The DNA wraps left-handedly 1.8 times around each nucleosome. The contribution to SLk is -1.8 for each nucleosome and 0 for each linker region. For the entire structure SLk = -9.

The linking number of the DNA on the relaxed SV40 minichromosome is measured in an indirect way. First, the DNA is stripped of the nucleosome particles, becoming in the process a plectonemically interwound free DNA. By means of gel electrophoresis, its linking number can be experimentally measured. In actuality, what is measured is the difference of its linking number and the linking number of the same DNA totally relaxed, ΔLk, as defined at the end of section 1. ΔLk is found to be about -1 per nucleosome core, i.e. ΔLk = -25 [8]. In section 1 we stated that relaxed SV40 has a linking number, Lk_0, of approximately 500. Therefore minichromosomal SV40 has Lk = 475.

We can now answer an important question: is the number of base pairs per turn, the helical repeat, unchanged from the 10.5 of relaxed DNA, when DNA is wrapped on the nucleosome? The answer must be negative because of the relationship Lk = SLk + Φ. Thus, we can theoretically determine, that since Lk = 475 and SLk = -45, that Φ must be 520. However, we have seen that Φ for relaxed SV40 is equal to Lk_0 = 500. Since Φ = 520 for minichromosomal SV40, the average helical repeat equals 5250/520 = 10.10. In this analysis, we have made a great many simplifications, but it is noteworthy that this number is in remarkably good agreement with the number 10.17 which is obtained by nuclease digestion experiments. Thus the answer to the question has been proven also by experiment to be negative.

To summarize, we have found a fundamental relationship Lk = SLk + Φ for three quantities that are directly accessible to experiment, Lk by electrophoresis, SLk by x-ray diffraction, and Φ by digestion. If two of the three are known, one can use the relationship to predict and therefore verify the experimental evidence of finding the third. This gives a powerful use of differential topology in the field of molecular biology.

Appendix

Definition: Let **A** be a closed differentiable curve and let **T** be its unit tangent vector field. Let **v** be a unit differentiable normal vector field along **A**. Let **C** be a closed curve which is obtained from **A** by a small displacement ε along **v**. ε should be so small that during the displacement no crossings of the displaced curve and **A** takes place. Then the *twist of C about A* is defined by the line integral:

$$\text{Tw}(\mathbf{C},\mathbf{A}) = \frac{1}{2\pi}\int_{\mathbf{A}} d\mathbf{v} \cdot \mathbf{T} \times \mathbf{v}$$

Theorem 1 [10]. *Let **A** be a closed oriented space curve and let v be a differential vector field which is normal to the curve. Let **C** be a closed curve which is obtained from **A** by a small displacement along v, such that during the displacement **C** does not cross **A**. Let Wr(**A**) be the writhing number of **A**, Tw(C,A) be the twist of C about **A**, and Lk(C,A) be the linking number of **C** with **A**. Then*

$$\text{Lk}(\mathbf{C},\mathbf{A}) = \text{Tw}(\mathbf{C},\mathbf{A}) + \text{Wr}(\mathbf{A}).$$

Theorem 2. *Let v_1 and v_2 be two differentiable vector fields along the curve A. Let C_1 and C_2 be two curves obtained by small displacements along these vector fields. Then,*

$$\text{Lk}(C_1,A) - \text{Lk}(C_2,A) = \text{Tw}(C_1,A) - \text{Tw}(C_2,A) = \Phi(C_1,C_2)$$

$\Phi(C_1.C_2)$ is the number of times that v_1 rotates past v_2, as the curve A is traversed.

Proof of Theorem 2: The first equality in the theorem follows immediately by subtracting the statement of Theorem 1 for the case of C_2 from that for C_1. The proof of the second of the equalities is as follows. We first observe that v_1 and v_2 are in the same plane that is perpendicular to the unit tangent vector T at each point. This plane is spanned by v_2 and $T \times v_2$. two mutually orthogonal unit vectors. Hence, v_1 is a linear combination of these two vectors. If ϕ is the angle between v_1 and v_2, then we may write:

$$v_1 = \cos(\phi)v_2 + \sin(\phi)T \times v_2$$

and hence,

$$T \times v_1 = \cos(\phi)T \times v_2 - \sin(\phi)v_2.$$

A simple computation yields:

$$dv_1 \cdot T \times v_1 = d\phi + dv_2 \cdot T \times v_2.$$

Integrating both sides of this last equation along the curve **A**, we obtain that:

$$\frac{1}{2\pi}\int_A dv_1 \cdot T \times v_1 = \frac{1}{2\pi}\int_A d\phi + \frac{1}{2\pi}\int_A dv_2 \cdot T \times v_2$$

or $Tw(\mathbf{C_1},\mathbf{A}) = \Phi(\mathbf{C_1},\mathbf{C_2}) + Tw(\mathbf{C_2},\mathbf{A})$, where $\Phi(\mathbf{C_1},\mathbf{C_2})$ measures the total turning of $\mathbf{v_1}$ about $\mathbf{v_2}$. It is the application of theorem 2 that gives the main result discussed in the text: that the linking number of a closed DNA on a protein surface is the sum of the surface linking number plus the winding number.

References

1. W.R. Bauer, F.H.C. Crick and J.H. White, Supercoiled DNA, Scientific American **243** (1980), 118-122.

2. T.C. Boles, J.H. White and N.R. Cozzarelli, Structure of plectonemically supercoiled DNA, J. Mol. Biol. **213** (1990), 931-951.

3. H.R. Drew and A.A. Travers, DNA bending and its relation to nucleosome positioning, J. Mol. Biol. **186** (1985), 773-790.

4. J.T. Finch, et al., X-ray diffraction study of a new crystal form of the nucleosome core showing higher resolution, J. Mol. Biol. **145** (1981), 757-769.

5. J.T. Finch, et al., Structure of nucleosome core particles of chromatin, Nature **269** (1977), 29-36.

6. F.M. Richards, Areas, volumes, packing and protein structure, Ann. Rev. Biophys. Bioeng. **6** (1977), 151.

7. T.J. Richmond, et al., Structure of the nucleosome core particle at 7 A resolution, Nature **311** (1984), 532-7.

8. M. Shure and J. Vinograd, The number of superhelical turns in native virion SV40 DNA and minicol DNA determined by the band counting method, Cell **8** (1976), 215-226.

9. J.M. Sogo, H. Stahl, T. Koller and R. Knippers, Structure of replicating simian virus 40 minichromosomes: The replication fork, core histone segregation and terminal structures, J. Mol. Biol. **189** (1986), 189-204.

10. J.H. White, Self-linking and the Gauss integral in higher dimensions, Am. J. Math. **91** (1969), 693-728.

11. J.H. White, *An introduction to the geometry and topology of DNA structure*, CRC Press, Inc., Boca Raton, 1989.

12. J.H. White and W.R. Bauer, Calculation of the twist and the writhe for representative models of DNA, J. Mol. Biol. **189** (1986), 329-341.

13. J.H. White and W.R. Bauer, Applications of the twist difference to DNA structural analysis, Proc. Natl. Acad. Sci. USA **85** (1988), 772-776.

14. J.H. White, N.R. Cozzarelli and W.R. Bauer, Helical repeat and linking number of surface wrapped DNA, Science **241** (1988), 323-327.

DEPARTMENT OF MATHEMATICS, UNIVERSITY OF CALIFONIA AT LOS ANGELES, LOS ANGELES, CA 90024-1555

E-mail: jhw@math.ucla.edu

Proceedings of Symposia in Applied Mathematics
Volume 45, 1992

Knot Theory and DNA

DE WITT L. SUMNERS

1. Introduction

Geometry and topology are subjects in the mainstream of pure mathematics. The word "pure" in the previous statement refers to the fact that the mathematics, although often originating in attempts to understand the "real" world, was subsequently abstracted from its origins and enjoys a vigorous internal intellectual life of its own, generating progress and excitement among its adherents. For example, knot theory (a subfield of topology) traces its mathematical origins to the 19th century work of Gauss, Listing, Helmholtz, Kelvin, Maxwell and Tait [Ar,Kn,T]. In the intervening century since the compilation of the first knot table, a great deal of progress has been made in "pure" knot theory. Knot theory is now enjoying a return to its mathematical "roots", in that the results of the century of mathematical progress are proving to be effective in the analysis of laboratory experiments in the natural sciences, and are generating high-level interactions between mathematics and physics. This article tells the story of the perhaps unexpected applicability of knot theory in the analysis of enzyme action on DNA.

In order to initiate and control various life processes of the cell such as replication, transcription and recombination, enzymes manipulate DNA in topologically interesting ways. These enzyme actions include promoting the coiling up (**supercoiling**) of DNA, passing one or more strands of DNA through a transient enzyme-bridged break in the DNA,

1991 Mathematics Subject Classification. Primary 57M25,92C40.
Research partially supported by the National Science Foundation.
This paper is in final form and no version of it will be submitted for publication elsewhere.

and breaking a pair of DNA strands apart and recombining them to different ends. The three-dimensional shape of these enzymes, the structure of the active enzyme-DNA complexes which are formed when the enzymes bind to the DNA, and the changes in the geometry and topology of DNA caused by enzyme action are of great biological interest. Experimental determination of the spatial conformation of DNA, proteins, and protein-DNA complexes in solution is difficult to do directly, and indirect methods are often used. One such method is the **topological approach to enzymology,** in which enzyme experiments on circular DNA substrate molecules are performed. The circular form of the DNA traps some of the enzyme-caused changes in DNA geometry and topology. By observing the changes in DNA geometry (supercoiling) and topology (knotting and linking), the enzyme mechanism can be described and quantized. These experiments are designed to exploit the descriptive and calculational abilities of geometry and topology. We will review some of these experimental results, and the mathematics that sustains their analysis. We will focus on the tangle model for site-specific recombination. This model uses experimental information (gel electrophoresis and electron microscopy) to write down tangle equations which quantize changes in DNA topology. The solution of these tangle equations involves some recently-developed knot theory. Using this model, the enzyme mechanism and the spatial structure of the active enzyme-DNA complex can be computed.

This paper will give a short, hopefully "user-friendly" introduction to the mathematics of knots, particularly those parts of the theory which are useful in molecular biology. Knot theory has an intuitive, hands-on, picture-driven nature which we hope to exploit as an aid to understanding. For more complete treatments of knot theory, see [CF,BZ,R, K1]. For more information on DNA applications of knot theory, see [ES1,S2,S3,SE,WC1]. This article is designed to provide an informal, intuitive introduction to the mathematical ideas of use in the application of knot theory to the analysis of DNA experiments. The list of references at the end of this paper contains some items in addition to those specifically referenced, articles which the reader may find of interest. For those readers who desire more mathematical detail, the material following the asterisk * at the end of each section may be of interest. This material can be omitted without harm to the continuity of the exposition.

2. Knot Theory

Topology is the mathematical science of shape, which aims to describe and quantize objects according to mathematically rigorous definitions of shape. Whereas geometry allows rigid transformation as its equivalence relation, topology allows non-rigid, flexible equivalence of objects. This flexible equivalence is advantageous in the application of topology to natural science. For example, consider the problem of the spatial description of macromolecules such as polymers and DNA. A single macromolecule is large and flexible. Such a molecule can assume a variety of configurations, driven from one to another by thermal motion, solvent effects, experimental manipulation, etc. Topology can help with the description and computation all of the possible configurations which can be attained from a given starting configuration. Of course, many of the configuration changes which are topologically feasible are physically impossible because of the unlimited flexibility of topological equivalence. Topology can, however, detect mathematical barriers which may exist between conformations, barriers which can only be crossed by the transient breaking and reunion of the macromolecules [Wa2,WC1]. For example, the helical Crick-Watson structure of duplex DNA imposes topological obstructions to life processes such as replication and transcription. It is the very existence of topological barriers which initiated the search for enzymes which can overcome them.

Knot theory (in 3-space) is the study of entanglement of flexible circles and arcs in R^3. Circles can become entangled with each other, forming **links**, and self-entangled, forming **knots**. Just as one disentangles an electrical extension cord by threading the ends through the entanglement in order to reduce it, an arc in 3-space cannot be entangled with itself, or with other arcs and circles. If, however, the arc is extremely long, or is periodically attached to other structures, entanglement (both transient and permanent) can occur.

Let R^n denote Euclidean space of dimension n, D^n the unit disk in R^n, and $S^{n-1} = \partial D^n$ the unit sphere in R^n which is the boundary of D^n. A **knot** is a finite subpolyhedron $k \subset R^3(S^3)$ which is homeomorphic to S^1. By considering only finite polygons, we avoid infinite complexity, such as that obtained by tying an infinite sequence of ever smaller knots converging to a point. An r-component **link** is the union of r pairwise disjoint knots $L = k_1 \cup ... \cup k_r$. A knot is then a link of one component, a single polygonal circle embedded in 3-space. Notation which helps to capture the essence of knotting is to denote a knot as a pair of spaces (R^3,k) or (S^3,k), a notation in which both the ambient space and the subspace are recognized. We note that as an alternative definition, one could require that the knot k be a smooth submanifold of the ambient space $R^3(S^3)$. In other words, we could choose to do knot theory in the smooth category, instead of the piecewise-linear (PL) category; it turns out that one obtains identical theories in the two categories, and it is intuitively more convenient to work in the PL category. On the other hand, all of our drawings of knots in this paper will be in the smooth category, where one can imagine that the corners on the knots have been rounded off, or are too small to be seen in the drawing. In the following discussion, unless otherwise specified, all maps will be PL. It is often mathematically more convenient to describe knots in the 3-sphere S^3, because S^3 is compact, and because a tamely embedded (finite polygonal) S^2 in S^3 separates S^3 into two 3-balls B^3, regions homeomorphic to D^3. It is however intuitively more convenient to think about knots in R^3. As a practical matter, since S^3 is the one-point compactification of R^3, the one extra point at infinity in S^3 makes no real difference to the theory. Unless otherwise stated, we will always take R^3 as our ambient space. In order to describe the flexible equivalence of knots and links, we must be careful about orientations. We take our ambient space R^3 to have a fixed (right-handed) orientation, where the right-hand thumb corresponds to the X-axis, the right-hand index finger corresponds to the Y-axis, and the right-hand middle finger corresponds to the Z-axis. One could shrink down and translate this XYZ frame at the origin to all points in the space, so that each point of R^3 comes equipped with a local orientation frame. A homeomorphism from R^3 to R^3 is **orientation preserving** if the local right-handed frame at each point of the domain maps to a local right-hand frame in the range. A reflection in a hyperplane (such as f(x,y,z) = (x,y,-z) which reflects in the XY plane) reverses the orientation of R^3. We might also require that the circular subspace k come equipped with an orientation (arrow). If so, we say that our knot k is **oriented**, if not, we say that our knot k is **unoriented**. Unless otherwise specified, all our knots will be **unoriented**. We say that two knots k_1 and k_2 are **equivalent** if there is an orientation-preserving (on R^3) homeomorphism of pairs $H:(R^3,k_1) \to (R^3,k_2)$. On a more intuitive level, two knots are equivalent if there is an elastic motion of the ambient space R^3 which moves k_1 into a position congruent to k_2. This elastic motion is a 1-parameter family of homeomorphisms (an **ambient isotopy**) of the pair (R^3,k_1) which starts with the identity and ends with H. For the proof that the existence of H implies the existence of this ambient isotopy, see [R]. If two knots are equivalent, then we say that the knots are of the same **knot type**, and that a given polyhedron k is a **representative** of its knot type. We will use the word "knot" to refer either to a representative or to an equivalence class of representatives, the usage to be determined by the context. Equivalence of links is defined similarly. For a link of $r \geq 2$ components, the link is **ordered** if the components are indexed by the integers 1,...,r. We might require that the homeomorphism on the ambient space preserve the ordering and the orientation of each component. Unless otherwise specified, our links will be unordered and unoriented; moreover, all of the links used to describe biological phenomena in this paper will be links of 2 components.

In order to describe a particular link, one usually considers a planar projection of k. The projection direction can be chosen so that no more than two strands cross at any point in the projection. We will take the XY plane in R^3 as the plane in which our link diagrams live. The crossings in a 2-dimensional projection can be coded by drawing the underpassing strand with a break at the crossing point, so that the link type can be uniquely reconstructed from it. Such a coded projection is called a **link diagram**. Figure 1 shows link diagrams; figure 1A is the **trivial knot** or **unknot,** the unique knot type which admits a diagram with no crossings. The **trivial link of r components** has a diagram with r circles and no crossings (not pictured).

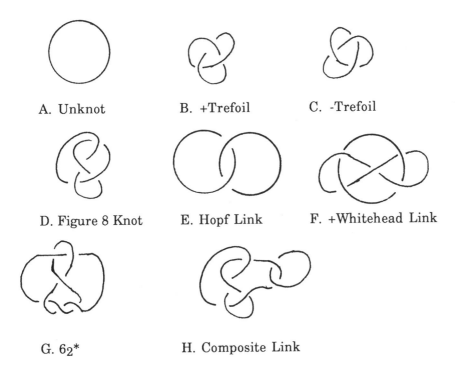

A. Unknot   B. +Trefoil   C. -Trefoil

D. Figure 8 Knot   E. Hopf Link   F. +Whitehead Link

G. 6_2^* H. Composite Link

Figure 1: Link Diagrams

A. +Crossing   B. -Crossing

Figure 2: Signed Crossing Convention

Given an oriented link diagram, one can assign a "+" or "-" sign to each of the crossings in the diagram using the convention shown in figure 2. This sign convention is determined by the right-hand rule for orientation of R^3, and can be easily remembered as follows: grasp the overcrossing string with your right hand, with your thumb in the direction of the overcrossing arrow. If the rest of your fingers point in the direction of the underpassing arrow, the crossing is a "+" crossing; if your fingers point in the opposite direction, the crossing is a "-" crossing. Note that for a knot diagram, either orientation of the knot determines the same sign at each of its crossings, because reversal of the orientation of the knot reverses the orientation of both strings involved in each crossing, so that the sign of each crossing is unchanged. The **+ trefoil** knot diagram of figure 1B has 3 "+" crossings, the **- trefoil** diagram of figure 1C has 3 "-" crossings, the **figure 8 knot** diagram of figure 1D has 2 "+" and 2 "-" crossings, and the knot **6_2^*** diagram (it is the mirror image of 6_2 in [R]) of figure 1G has 4 "+" and 2 "-" crossings. For an oriented link diagram, given a crossing formed by different link components, reversing the orientation of one of the components reverses the sign of that crossing. Given a 2-component oriented link diagram for the link $L = k_1 \cup k_2$, the **linking number $Lk(k_1,k_2)$** is defined as half of the sum of the signed crossings, where the summation is performed over all the crossings of different components; self-crossings of components are ignored. For example, if each component of the **Hopf link** of figure 1E is oriented with an arrow which points toward the top of the page, the linking number is +1; if the components of the **+Whitehead link** of figure 1F are given any orientation, the linking number is zero; the linking number of the link of figure 1H is ± 1, depending on orientations.

The effect of reversing the orientation of R^3 on a link can be seen in a diagram of the link. If the plane of the diagram is the XY plane and the Z axis points out of the page, reversing the orientation of space by reflection in the XY plane reverses the role of over and under at each crossing, and one obtains a diagram for the "mirror image" of a link. For example, the +trefoil and the -trefoil in figure 1 are mirror images of each other. If a link is not equivalent to its mirror image, it is said to be **chiral**; otherwise the link is **achiral**. For example, the unknot (figure 1A), the figure 8 knot (figure 1D), the (unoriented) Hopf link (figure 1E) and the link of figure 1H are achiral. The two trefoils (figures 1B and 1C) are inequivalent and form a chiral pair of mirror image knots; the (unoriented) +Whitehead link (figure 1F) and the knot $6_2{}^*$ (figure 1G) are chiral (mirror images are not shown). The knots and links in figures 1B-G are examples of **prime** links--they cannot be decomposed into simpler knots and links. Trivial links are (by definition) not prime. Figure 1H is a diagram of a **composite** link. More precisely, a link (S^3, L) is prime if every tame 2-sphere in S^3 which intersects the link in 2 points has the property that exactly one of the two 3-balls into which S^3 is divided by the 2-sphere intersects L in an unknotted arc, and given any tame 2-sphere which does not intersect the link, all of the link components are contained in the same 3-ball in the complement of the 2-sphere.

The problem with knot(link) diagrams is that each knot(link) type admits infinitely many diagrams. Given a fixed polyhedral representative of a knot, one could rigidly rotate it and then project it, obtaining a number of different diagrams; or, one could non-rigidly move the knot about in space before projecting it, obtaining (if one so desires) a diagram with many extraneous crossings. How then, given a pair of diagrams, is one to decide whether or not they represent the same knot(link)? This is a difficult question. As is explained in the article of Kauffman in this volume [K], two diagrams represent the same link if and only if there is a finite sequence of Reidemeister moves which start with one of the diagrams and convert it to the other diagram. One can use the theorem of Reidemeister (and other tools of topology) to define topological invariants, numbers, polynomials, etc. which can unambiguously be assigned to a knot type. These invariants can usually be computed directly from any diagram for the knot(link). For example, the linking number of a 2-component link is an invariant of oriented link type. If two links differ in any invariant, then they are inequivalent. On the other hand, if no invariant resolves them, then they must either be proved identical by hands-on manipulation, or resolved by some new invariant.

For a given link type, one usually selects a diagram of minimum complexity, one with the fewest crossings possible. This minimum number of crossings is the **crossing number** of the link, and a diagram which exhibits the minimum number of crossings is a **minimal diagram**. All of the diagrams in figure 1 are minimal diagrams. The diagrams in figure 1A-G are **alternating diagrams**, having the property that as one traverses any component, one alternately encounters over- and under- passes. Since there are only finitely many diagrams of n crossings, there are only finitely many knots and links of n or fewer crossings. One usually catalogs knots and links by their crossing number, obtaining knot and link tables [R]. Knot and link tables generally list only prime knots and only one representative from each chiral pair of knots(links).

3. 4-plats

We will now describe the family of 4-plats, a family of prime knots and 2-component links which have three distinct advantages for biological application: (1) These knots and links are formed by twisting pairs of strings about each other (**plectonemic supercoiling**), and mimic the supercoiling one sees in electron micrographs of DNA. (2) All prime knots of 7 or fewer crossings are 4-plats; all prime links of 6 or fewer crossings are 4-plats. So all small crossing number (prime) knots and links are 4-plats. Many of the DNA knots and links

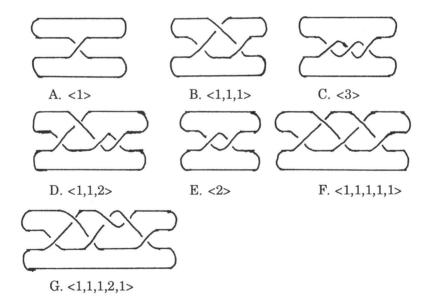

A. <1>   B. <1,1,1>   C. <3>

D. <1,1,2>   E. <2>   F. <1,1,1,1,1>

G. <1,1,1,2,1>

Figure 3: Canonical 4-plat Diagrams

observed in the laboratory are 4-plats because the DNA substrates tend to be of low molecular weight, having on the average 5,000-10,000 base pairs. Duplex DNA is fairly stiff, and it is energetically difficult to tie large crossing number knots in short, stiff DNA. (3) The class of 4-plats is mathematically tractable, being completely classified by a system of vectors which are of use in computations which arise in the analysis of DNA experiments.

A **4-plat** (**2-bridge**) knot or 2-component link is one which admits a canonical diagram which consists of a braid on four strings, in which only the top three strings are involved in crossings, and the bottom string is crossing free [BZ]. These knots and links also admit other (nonminimal and nonalternating) diagrams with 2 (maximal) overcrossings, hence the designation "2-bridge". Figure 3A-G shows the canonical 4-plat diagrams for the knots and links of figure 1A-G. The link of figure 1H has no 4-plat diagram, since all non-trivial 4-plats are prime [BZ]. The pattern of crossings in a canonical 4-plat diagram can be represented by an odd-length vector $<c_1, \ldots, c_{2k+1}>$ with positive integer entries. The correspondence between the 4-plat diagram and the vector is obtained as follows: Number the string positions in the 4-plat from 1-4 from top to bottom. Beginning at the left-hand end of the diagram, the string in position 3 crosses over the string in position 2 c_1 times, forming a horizontal row of left-handed crossings (plectonemic intertwines). Next, the string in position 1 crosses over the string in position 2 c_2 times, forming a horizontal row of right-handed intertwines (the same handedness as the usual Crick-Watson structure of duplex DNA). We continue, alternating between handedness and string position, terminating with c_{2k+1} left-handed crossings between the strings in positions 2 and 3 at the right-hand end of the diagram. The vector which corresponds to a canonical 4-plat diagram is called a **classifying vector** for the 4-plat. Canonical 4-plat diagrams represent either knots or 2-component links; moreover, the 2-component links have the property that each of the components is unknotted. Moreover, canonical 4-plat diagrams are alternating, and are minimal [ES], with the single exception of $<1>$ which represents the trivial knot. The convention that the entries of the vector be positive excludes the vector $<0>$, which corresponds to the 4-plat diagram with no crossings (the trivial link of 2 components), so we add this classifying vector to our list of classifying vectors. If the canonical 4-plat diagram with classifying vector $<c_1, \ldots, c_{2k+1}>$ is rotated by 180 degrees about a vertical axis in the plane of the diagram, one obtains another canonical diagram with classifying vector $<c_{2k+1}, \ldots, c_1>$, the reverse of the original vector. Since the link type is unchanged by this rotation, both vectors represent the same link type. This is the only ambiguity in the classification scheme given by the vectors: two canonical 4-plat diagrams represent the same knot (link) type iff they have identical vectors, or their vectors become identical if one of them is reversed.

*If one relaxes the condition on the classifying vector to allow the possibility of zero or negative integer entries, to each odd-length vector

$<d_1, ... ,d_{2k+1}>$ with integer entries there corresponds a 4-plat diagram, but in general this diagram is non-canonical, neither alternating nor minimal. Nevertheless, from any vector representative $<d_1, ... ,d_{2k+1}>$ for the 4-plat, it is possible to calculate a classifying extended rational number $\beta/\alpha \in \{\mathbb{Q} \cup \infty\}$ as a continued fraction: $\beta/\alpha = 1/(d_1+(1/d_2+...))$. If one performs this continued fraction calculation on a classifying vector for the 4-plat k (k $\neq$ <0>, <1>), one obtains $0 < \beta < \alpha$. Unless otherwise specified in the following, we will always choose to compute a classifying rational number for the 4-plat from a classifying vector, and (following [BZ]) we write the 4-plat as $b(\alpha,\beta)$. The numbers α and β have a geometric interpretation in terms of a 2-bridge projection of a 4-plat (see [BZ,p183]). With the crossover convention of Figure 3, the 2-fold branched cyclic cover of $b(\alpha,\beta)$ is the lens space $L(\alpha,\beta)$. For example, the unknot b(1,1) = <1> has S^3 as 2-fold branched cover, and the trivial link of 2 unknotted components b(0,1) = <0> has $S^1 \times S^2$ as 2-fold branched cover. 4-plats are classified by means of their 2-fold branched cyclic covers :

THEOREM 3.1 [BZ]: *Two 4-plats $b(\alpha,\beta)$ and $b(\alpha',\beta')$ are equivalent (as unoriented knots or links) iff $\alpha = \alpha'$ and $\beta^{\pm 1} \equiv \beta'$ (mod α)* .

4. Rational Tangles

We will now describe a mathematical configuration of a pair of arcs in a 3-ball, called a tangle. The mathematical notion of tangles is due to John Conway [C]. We will focus on the family of rational tangles, which have the following advantages for biological applications: (1) Like 4-plats, these configurations are formed by the intertwining of pairs of strings (plectonemic supercoiling), and mimic DNA electron micrographs. (2) We can use tangles to model the enzyme-DNA complex, with the enzyme being the 3-ball, and the 2 strings being the two sites at which the enzyme is attached to the DNA. (3) Rational tangles can be added together, and form building blocks for more complicated tangles and for 4-plats. (4) Rational tangles, like 4-plats, are completely classified by vectors, and the classification scheme can be exploited to solve tangle equations which arise in DNA experiments.

Consider D^3, the unit 3-ball in R^3. The **equator** of D^3 is the circle of intersection of the $S^2 = \partial D^3$ with the XY plane; the **equatorial D^2** of this 3-ball is the intersection of this 3-ball with the XY plane. On the unit 3-ball, select 4 points on the equator (called NW, SW, SE, NE). A (**2-string) tangle** in D^3 is a configuration of two disjoint strings in the unit 3-ball whose endpoints are the 4 special points (NW, SW, SE, NE). Two tangles in D^3 are **equivalent** if it is possible to elastically transform (inside the confining D^3) the strings of one tangle into the strings of the other, without moving the endpoints {NW,SW,SE,NE} and without breaking a string or passing one string through another. A class of equivalent tangles is called a **tangle type**. Tangle theory is knot theory done inside D^3, with the ends of the strings firmly glued down. Tangles are represented by projecting them onto the equatorial D^2 in D^3, obtaining tangle diagrams, as in figure 4. Two tangle diagrams represent equivalent tangles iff one diagram can be converted to the other by a finite sequence of Reidemeister moves in the interior of D^2. In all figures containing tangles we assume that the 4 boundary points {NW, SW, SE, NE} are as in figure 4A, and we suppress these labels.

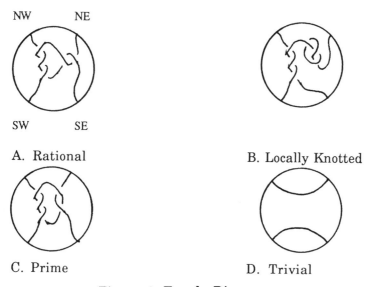

A. Rational   B. Locally Knotted

C. Prime   D. Trivial

Figure 4: Tangle Diagrams

All four of the tangles in figure 4 are pairwise inequivalent. However, if we relax the restriction that the endpoints of the strings remain fixed, and allow the endpoints of the strings to move about on the surface 2-sphere of the 3-ball, then the tangle of figure 4A can be transformed into the trivial tangle of figure 4D. The tangles in figures 4B and 4C cannot be transformed to the trivial tangle by any sequence of such twisting motions of the endpoints on S^2. The family of tangles which can be converted to the trivial tangle by twisting the endpoints of the strings on the boundary 2-sphere is the family of **rational tangles**. Equivalently, a rational tangle is one in which the strings can be deformed (leaving the endpoints fixed) into $S^2 = \partial D^3$. The tangle of figure 4B is **locally knotted**, and the tangle of figure 4C is **prime**. Every tangle falls into one of the three classes {rational, prime, locally knotted} [L2].

In order to compare tangles in different balls in R^3, we need to think of them as having "the same" boundary. Let (B^3,t) represent a pair of strings properly embedded in an oriented 3-ball. As in [BS], we define the **model 2-sphere** S^2 in R^3 to be the boundary of D^3, equipped with 4 distinguished equatorial points P = {NE, SE, SW, NW}. We require that every tangle (B^3,t) comes equipped with a **boundary parameterization**, that is, a homeomorphism $\Phi: (\partial B, \partial t) \to (S^2,P)$. So, a tangle is a triple $B = (B,t,\Phi)$. Two tangles (B,t,Φ) and (B',t',Φ') are **equivalent** if there is an orientation-preserving homeomorphism H: $(B,t) \to (B',t')$ such that $\Phi = \Phi'H$ on ∂B.

Rational tangles form a homologous family of 2-string configurations in a confined region of 3-space, and are formed by a pattern of plectonemic supercoiling of pairs of strings. Like that for 4-plats, there is a classification scheme for rational tangles which corresponds to a canonical minimal alternating diagram. The classifying vector for a rational tangle is an integer-entry vector $(a_1,...,a_n)$ of odd or even length, with all entries (except possibly the last) non-zero and having the same sign, and with $|a_1| > 1$. The integers in the classifying vector represent the left-to-right (west to east) alternation of vertical and horizontal windings in the standard tangle diagram, always ending with horizontal windings on the east side of the diagram. Horizontal winding is that between strings in the north and south positions; vertical winding is that between strings in the west and east positions. By convention, positive integers correspond to horizontal right-handed supercoils and vertical left-handed supercoils, and negative integers correspond to horizontal left-handed supercoils and vertical right-handed supercoils. A tangle classified by a vector of length one is called an **integral tangle**. Figure 5 shows some canonical rational tangle diagrams. Two rational tangles are of the same type iff they have identical classifying vectors. Due to the requirement that $|a_1| > 1$ in the classifying vector convention for rational tangles, the corresponding tangle projection must have at least 2 crossings. There are 4 rational tangles {(0),(0,0),(1),(-1)} which are exceptions to this convention, and are displayed in figure 5C-F. Analogous to the situation for 4-plats, from any vector representative $(b_1, ... ,b_n)$ for the tangle, one can calculate a classifying extended rational number $\beta/\alpha \in \{\mathbb{Q} \cup \infty\}$ as a continued fraction: $\beta/\alpha = b_n + 1/(b_{n-1} + (1/(b_{n-2} + ...)))$.

The tangle (0,0) corresponds to $1/0 = \infty$ in this scheme. Two rational tangles are of the same type iff these (extended) rational numbers are equal [C], which is the reason for calling them "rational" tangles. We denote a rational tangle X by its classifying vector $X = (a_1,...,a_n)$, or by its classifying extended rational number $X = \beta/\alpha$. For any tangle X, the string which begins at the NW position in a tangle must terminate in one of three places, namely {NE,SE,SW}. Accordingly, we say that if the termination point is NE, we say that X has the **parity of (0)**, since (0) is the simplest tangle with this property. Likewise, if the string terminates at SE, we say that X has the **parity of (1)**; and if the string terminates at SW, we say that X has the **parity of (0,0)**.

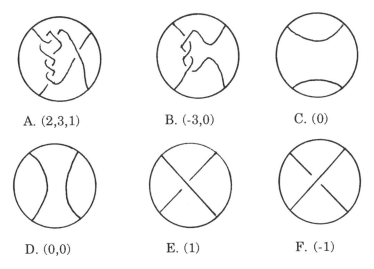

A. (2,3,1)   B. (-3,0)   C. (0)

D. (0,0)   E. (1)   F. (-1)

Figure 5: Rational Tangles

In order to use tangles as building blocks for knots and links, we now introduce the geometric operations of tangle addition and tangle closure. Given tangles X and Y, one can form the **tangle sum** $(X+Y)$ as shown in figure 6A. Tangle addition is associative but not commutative. The sum of two tangles may not always be a tangle. For example, if both X and Y are of parity (0,0), then $X+Y$ is not a 2-string tangle because it contains a circle in addition to two arcs. If the sum of two rational tangles is a tangle, it may not be a rational tangle. Given any tangle Z, one can form the closure $N(Z)$ as in figure 8B. In the closure operation on a 2-string tangle, ends NW and NE are connected, and ends SW and SE are connected, and the defining ball is deleted, leaving a knot or a link of 2 components. Deletion of the defining ball for a tangle corresponds to the biological event of deproteinization, when the enzyme and DNA are separated. One can combine the operations of tangle addition and tangle closure to form tangle equations of the form $N(X + Y)$ = knot(link); the tangles $\{X, Y\}$ are said to be **summands** of the resulting knot(link). An example of this phenomenon is the tangle equation $N((-3,0) + (1)) = <2>$, shown in figure 6C. In general, if X and Y are any two rational tangles, then $N(X+Y)$ is a 4-plat. Given these constructions, rational tangles are summands for 4-plats.

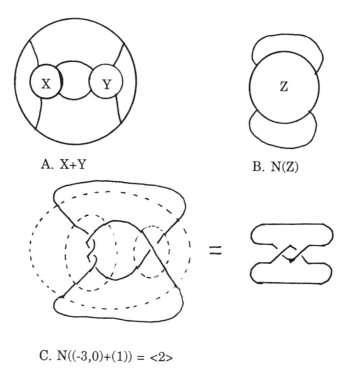

A. X+Y

B. N(Z)

C. $N((-3,0)+(1)) = <2>$

Figure 6: Tangle Operations

*Using the classifying vectors, one can develop a **tangle calculus** [ES1] to solve equations of the form $N(X+Y) = L$, where X and Y are tangles, and L is a knot or link two components. If both A and B are rational tangles, and $N(A + B) = L$, then L is a 4-plat. In fact, if $A = \alpha_1/\beta_1$ and $B = \alpha_2/\beta_2$, and $N(A+B) = L = b(\alpha, \beta)$, then $\alpha = |\alpha_1\beta_2 + \alpha_2\beta_1|$ [ES1]. Suppose, however, that A is a rational tangle, K is a 4-plat, and that we wish to solve the equation $N(X + A) = K$ for the unknown tangle X. Unfortunately, this data does not force X to be a rational tangle. For example, if $A = (n)$ (a horizontal row of n half-twists), and $K = <1>$ the unknot, the equation $N(X+(n)) = <1>$ has infinitely many distinct prime tangle solutions of the form $(B + (-n))$, where B denotes any prime tangle with the property that $N(B + (0)) =$ unknot. One such prime tangle is shown in figure 4C. Since B is prime, then $(B + (-n))$ is prime [L2], and $N((B + (-n)) + (n)) = N(B + (0)) = <1>$. Although prime tangle solutions to these equations are in general difficult to enumerate, all rational tangle solutions to equations of this type are given by the following theorem:

THEOREM 4.1 [ES1]: *Let $A = \beta / \alpha = (a_1, \dots, a_{2n})$ be a rational tangle, and $K = <c_1, \dots, c_{2k+1}> \neq <0>$ be a 4-plat. The rational tangle solutions to the equation $N(X + A) = K$ are the following:*
$X = (c_1, \dots, c_{2k+1}, r, -a_1, \dots, -a_{2n})$, or $X = (c_{2k+1}, \dots, c_1, r, -a_1, \dots, -a_{2n},)$,
r any integer.
If $K = <0>$, then $X = (-a_1, \dots, -a_{2n})$ is the unique solution.

In terms of classifying rational numbers, the rational solutions for X in the above theorem are given by the fractions:
$u/v = (\beta q - \alpha'p + r\beta p)/(\beta'p - \alpha q - r\alpha p)$, and
$u/v = (\beta q' - \alpha'p + r\beta p)/(\beta'p - \alpha q' - r\alpha p)$.

Although theorem 4.1 says that one equation in one unknown has infinitely many rational solutions ($K \neq <0>$), the next result says that two equations in one unknown has at most two rational solutions.

THEOREM 4.2 [ES1]: *Let $A_1 \neq A_2$ be rational tangles, and K_1 and K_2 be 4-plats. There are at most two distinct rational tangle solutions to the equations*
  (i) $N(X + A_1) = K_1$
  (ii) $N(X + A_2) = K_2$.

Proof:
     Let $X = u/v$, $A_1 = \beta_1/\alpha_1$, $A_2 = \beta_2/\alpha_2$, $K_1 = b(\alpha,\beta)$, and $K_2 = b(\alpha',\beta')$. Then we have $\alpha = |\ u\alpha_1 + v\beta_1\ |$ and $\alpha' = |\ u\alpha_2 + v\beta_2\ |$. In the (u,v)-plane, these equations describe two pairs of parallel straight lines. These lines intersect in at most 4 points. Since $u/v = -u/-v$, these 4 points of intersection describe at most two distinct rational tangle solutions for the equations in the hypothesis.

5. The Topological Approach to Enzymology

We will now consider the situation of enzymes operating on covalently closed circular duplex DNA. Duplex DNA consists of two linear backbones of sugar and phosphorus. Attached to each sugar is one of the four bases: A=Adenine, T=Thymine, C=Cytosine, G=Guanine. The base compounds form hydrogen bonds with each other to form **base pairs**, where A binds only with T (and vice versa), and C binds only with G (and vice versa). If one reads along either of the backbone strands of a DNA molecule, one obtains a word in the letters {A,C,G,T}, called the **genetic sequence** of the DNA. Reading along the other backbone produces the "dual" sequence, with A and T interchanged, and C and G interchanged. As explained in the article of J.H. White [Wh], one can model duplex DNA as a ruled ribbon surface, the ruling arcs formed by hydrogen bonds between base pairs. In the classical Crick-Watson model for DNA, the ribbon surface is twisted in a right-handed helical fashion. If the axis of a DNA ribbon is planar (**relaxed DNA**), the pitch of the twisted ribbon is approximately 10.5 base pairs per full helical twist. If a pair of DNA ribbons are twisted about each other in a helical fashion (like the twisting of the DNA backbones about each other), each half twist forms a DNA **supercoil** [BCW]. If the axis of the DNA is non-planar, or the DNA is under stress, or bound to a protein, the helical pitch can change. Duplex DNA can exist in closed circular form, where the ribbon surface forms a twisted orientable band (instead of a twisted Mobius band), and the boundary of the band consists of two backbone circles. Given a circular duplex DNA molecule, one can "nick" (break) one of the two DNA backbones with an enzyme called **DNAse**, and the nicked molecule then loses its desire to supercoil in 3-space and "relaxes". Nicking one of the backbone strands does not change the knot type of the axis of the molecule. Relaxed unknotted circular duplex DNA is believed to assume a configuration in which its axis is nearly planar. After relaxation by nicking, one can then repair the nick in the DNA backbone with another enzyme called **ligase**, obtaining a relaxed DNA molecule with no breaks in either backbone. If one orients the backbone strands in a parallel fashion, the relaxed-state linking number (Lk_0) of the backbone strands of the ligated relaxed molecule can be used as a reference for measuring the effect of **topoisomerase** enzymes on DNA. These enzymes can pass backbone strands through each other via an enzyme-bridged transient break. When backbone strands are passed through each other on a circular duplex DNA molecule, the linking number of the backbone strands takes on a new value Lk. For unknotted circular duplex DNA, the change in linking number due to enzyme action $\Delta Lk = (Lk_0 - Lk)$ is an observable via gel electrophoresis, because a change of ± 1 in the linking number is converted by the local stiffness of the ribbon (its desire to maintain a locally constant helical pitch) into ± 1 supercoils [Wh]. The differential geometry of ribbons in 3-space plays an important role in understanding supercoil formation, helical twist of duplex DNA, and the mechanism of topoisomerase enzymes which pass DNA through itself in order to reduce molecular entanglement [WC1, S2].

The experimental strategy is to observe the enzyme-caused changes in the euclidean geometry (supercoiling) and topology (knotting and linking) of the DNA, and to deduce enzyme mechanism from these changes. This has been called the **topological approach to enzymology** [WC1], and is schematically depicted in figure 7. In figure 7 and all subsequent figures where DNA is depicted, only the axis of the duplex DNA molecule is drawn; the primary Crick-Watson helical structure is not shown, and supercoiling is often omitted. The geometry(supercoiling) and topology(knotting and linking) of the circular substrate are experimental control variables. The geometry and topology of the enzyme reaction products are observables. In figure 7, we start with an unknotted substrate molecule with one negative supercoil. (The sign of the supercoil is the sign of the crossing in the diagram.) Figure 7 shows a spectrum of possible products, ranging from an unknotted molecule with 2 negative supercoils (a change in supercoiling) to a trefoil knot (a change in knotting), to a Hopf link (a change in linking).

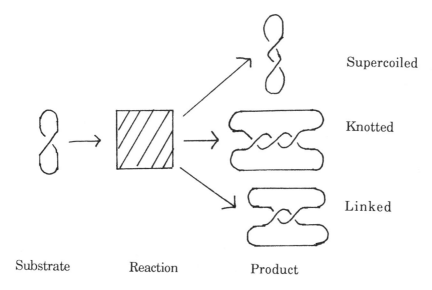

Figure 7: The Topological Approach to Enzymology

It is a new observation technique (*rec A* enhanced electron microscopy) [KS] which makes possible the detailed knot-theoretic analysis of reaction products. ***Rec A*** is an ***E. coli*** protein which binds to DNA, and whose function **in vivo** (in the cell) is to facilitate generalized recombination. Naked duplex DNA is approximately 20 angstroms in diameter, and *rec A* coated DNA is approximately 100 angstroms in diameter. The process of *rec A* coating fattens, stiffens and stretches (untwists) the DNA. The enlarged diameter of *rec A* coated DNA means that crossovers in electron micrographs of *rec A* coated DNA circles can (usually) be unambiguously resolved. **In vitro** (in the laboratory) experiments usually proceed as follows: A collection of artificial circular DNA substrate molecules is prepared by cloning techniques, with all of the substrate molecules representing the same knot type (usually the unknot). The genetic sequence of the substrate molecules can be controlled, and is arranged to suit the requirements of a particular experiment. The amount of supercoiling of the substrate molecules (the **supercoiling density**) is also a control variable. The substrate molecules are reacted with purified enzyme collected from living cells, and the reaction products are fractionated by **gel electrophoresis**. DNA molecules are naturally negatively charged, with the amount of negative charge proportional to the molecular weight. A gel is an obstructive protein medium through which the DNA molecules can be forced under the influence of an electric field. A DNA sample is placed at the top of a gel column, and similar molecules migrate in the electric field with similar velocities, forming discrete DNA bands in the gel when the electric field is turned off. The velocity of a molecule through the gel is a function of both the molecular weight and the (average) shape. For two molecules of the same molecular weight (as is the case in these experiments), the difference in gel velocity is due to differences in the geometry (supercoiling) and topology (knot and link type) of the DNA molecules. For example, in unknotted DNA, gel electrophoresis discriminates on the basis of number of supercoils, and can detect a difference of one in the number of supercoils. In gel electrophoresis of knotted and linked DNA, one must nick the reaction products prior to electrophoresis in order to relax the molecular knots and links, because supercoiling confounds the gel behavior. For nicked DNA knots and links, under the proper conditions gel velocity is (surprisingly) determined by the crossover number of the DNA knot or link; knots and links of the same crossover number migrate with approximately the same gel velocities [DS]. After running the gel, the gel bands are excised, and the DNA molecules are removed from the gel, and coated with *recA* protein. The DNA molecules are then placed on a grid and shadowed with platinum for viewing under the electron microscope. Electron micrographs of the reaction products (fig. 12) are made, and frequency distributions of knot and link types of the products are prepared.

Site-specific recombination is one of the ways nature alters the genetic code of an organism, either by moving a block of DNA to another position on the molecule (a move performed by **transposase**), or by integrating a block of viral DNA into a host genome (a move preformed by **integrase**). An enzyme which mediates site-specific recombination on DNA is called a **recombinase**. A **recombination site** for a given recombinase is a short (10-15 base pairs) linear segment of DNA whose genetic sequence is recognized by the recombinase. Site-specific recombination can occur when a pair of sites (on the same or on different DNA molecules) becomes juxtaposed in the presence of the recombinase. The pair of recombination sites is aligned (brought close together), probably through enzyme manipulation or random thermal motion (or both), and both sites (and perhaps some contiguous DNA) are then bound by the enzyme. This stage of the reaction is called **synapsis,** and we will call the protein-DNA complex (in the biochemical sense) formed by the bound DNA and the enzyme the **synaptosome**. We will call the entire DNA molecule(s) involved in synapsis (which includes the parts of the DNA molecule(s) not bound to the enzyme) together with the bound enzyme the **synaptic complex**. After forming the synaptosome, the enzyme then performs two double-stranded breaks at the sites, and recombines the ends by exchanging them in an enzyme-specific manner. The synaptosome then dissociates, and the DNA is released by the enzyme. We call the pre-recombination unbound DNA molecule(s) the **substrate**, and the post-recombination unbound DNA molecule(s) the **product**. The effect of a single recombination event on the genetic sequence near the recombination sites is shown in Figure 8. In these figures, duplex DNA is represented by a single line, and supercoiling is omitted. The recombination sites are represented as arrows, since the site sequence is not (in general) palindromic. Reading figure 8 left-to-right, one of the pre-recombination sites is depicted as a hollow arrow, the other as a solid arrow. The schematic action of recombination is to break each of the arrows just below the head, and attach the tail of each arrow to the head of the other arrow.

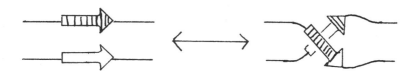

Figure 8: A Recombination Event

The process of recombination involves some interesting topological changes in the substrate. In order to trap these topological changes, one chooses to perform experiments on circular DNA substrate, because the topological effects of recombination on linear substrate would not be observable. One must perform an experiment on a large number of circular molecules in order to obtain a workable amount of product. Using cloning techniques, one can synthesize circular duplex DNA molecules which contain two copies of a recombination site. The orientation of each of the recombination sites induces an orientation on the ambient circular DNA molecule. If these induced orientations agree, this site configuration is called **direct repeats**, and if the induced

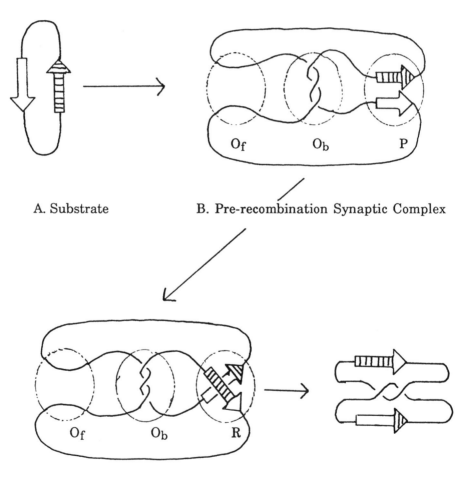

A. Substrate   B. Pre-recombination Synaptic Complex

C. Post-recombination Synaptic Complex   D. Product

Figure 9: Recombination With Direct Repeats

orientations disagree, this site configuration is called **inverted repeats**. If the substrate is a single DNA circle with direct repeats, the recombination product is a pair of DNA circles, and can form a DNA link (or **catenane**) (fig. 9). If the substrate is a pair of DNA circles with one site each, the product is a single DNA circle (fig. 9 read in reverse), and can form a DNA knot. If the substrate is a single DNA circle with inverted repeats, the product is a single DNA circle, and can form a DNA knot (fig. 10). **Distributive recombination** may occur, in which multiple recombination events occur during multiple binding encounters between enzyme and DNA. Under some conditions, however, **processive recombination** occurs, whereby the enzyme carries

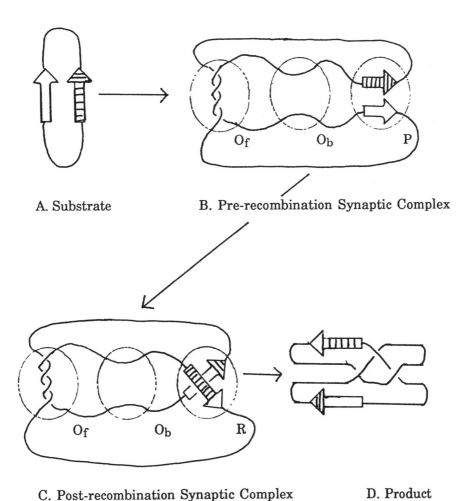

A. Substrate   B. Pre-recombination Synaptic Complex

C. Post-recombination Synaptic Complex   D. Product

Figure 10: Recombination With Inverted Repeats

out more than one strand exchange at a single binding encounter between enzyme and DNA.

In electron micrographs of synaptic complexes, the synaptosome is seen as a small black mass attached to supercoiled DNA. Fig. 11 is an electron micrograph of the synaptic complex formed by the enzyme **Tn3 resolvase** bound to circular DNA substrate with directly repeated sites. As is evident from figure 11, the synaptosome structure is difficult to observe by electron microscopy. We intend to use mathematics to compute the structure of the pre- and post-recombination synaptosome.

Figure 11: Resolvase Synaptic Complex (Courtesy of N.R. Cozzarelli)

6. The Tangle Model

The aim of our mathematical model is, given the observed changes in geometry and topology of the DNA, to compute the topology of the entire synaptic complex, both before and after enzyme action. Within the region controlled by the enzyme, the enzyme breaks the DNA at each site, and recombines the ends by exchanging them. We model the enzyme itself as a 3-ball. The synaptosome consisting of the enzyme and bound DNA forms a 2-string tangle.

What follows is a list of biological and mathematical assumptions made in the tangle model [ES1,S2,SE]. Most of these assumptions are implicit in the existing analyses of the results of enzyme experiments on circular DNA[CK,SSB,SSC,WD,KK,WM].

We make the following biological assumption:

BIOLOGICAL ASSUMPTION 6.1: *The enzyme mechanism in a single recombination event is constant, independent of the geometry (supercoiling) and topology (knotting and catenation) of the substrate population. Moreover, recombination takes place entirely within the domain of the enzyme ball, and the substrate configuration outside the enzyme ball remains fixed while the strands are being broken and recombined inside and on the boundary of the enzyme.*

That is, we assume that any two pre-recombination copies of the synaptosome are identical, meaning that we can by rotation and translation superimpose one copy on the other, with the congruence so achieved respecting the structure of both the protein and the DNA. We likewise assume that all of the copies of the post-recombination synaptosome are identical.

In a recombination event, we can mathematically divide the DNA involved into 3 types: (1) the DNA at and near the sites where the DNA breakage and reunion is taking place; (2) other DNA bound to the enzyme which is unchanged during a recombination event; and (3) the DNA in the synaptic complex which is not bound to the enzyme and which does not change during recombination. We make the following mathematical assumption about DNA types (1) and (2):

MATHEMATICAL ASSUMPTION 6.2: *The synaptosome is a 2-string tangle, and can be mathematically subdivided into the sum of two tangles.*

One tangle, the **parental tangle P**, contains the recombination sites where breakage and reunion takes place. The other tangle, the **outside bound tangle O_b**, is the remaining DNA in the synaptosome outside the P tangle; this is the DNA which is bound to the enzyme, but which remains unchanged during recombination. The **enzyme mechanism** is modeled as tangle replacement (surgery) in which the parental tangle P is removed from the synaptosome and replaced by the **recombinant tangle R**.

Therefore, our model assumes the following:

$$\text{pre-recombination synaptosome} = (O_b + P)$$

$$\text{post-recombination synaptosome} = (O_b + R).$$

In order to accommodate DNA of type (3), we let the **outside free tangle O_f** denote the synaptic complex DNA which is free (not bound to the enzyme), and which is unchanged during a single recombination event. We make the following mathematical assumption:

MATHEMATICAL ASSUMPTION 6.3: *The entire synaptic complex is obtained from the tangle sum   (O_f + synaptosome) by the numerator construction.*

If one deproteinizes the pre-recombination synaptic complex, one obtains the substrate; deproteinization of the post-recombination synaptic complex yields the product. The topological structure (knot and catenane types) of the substrate and product yield equations in the recombination variables $\{O_f, O_b, P, R\}$. Specifically, a single recombination event on a single circular substrate molecule produces two **recombination equations** in four unknowns:

(6.4)   Substrate equation: $N(O_f + O_b + P)$ = Substrate

(6.5)   Product equation: $N(O_f + O_b + R)$ = Product .

The geometric meaning of these recombination equations is illustrated in figures 9 and 10. In figure 9, O_f = (0), O_b = (-3,0), P = (0) and R = (1). With these values for the variables, our recombination equations become:

substrate equation: $N((0)+(-3,0)+(0))$ = <1>
product equation: $N((0)+(-3,0)+(1))$ = <2> .

In figure 10, we have O_f = (-4,0), O_b = (0), P = (0) and R = (1). With these values for the variables, our recombination equations become:

substrate equation: $N((-4,0)+(0)+(0))$ = <1>
product equation: $N((-4,0)+(0)+(1))$ = <1,1,1>

In order to solve the recombination equations for $\{O_f, O_b, P, R\}$, we need experimental information. Let the **outside tangle O** = $(O_f + O_b)$. The change in topology from substrate to product can be used to determine the tangle O; in order to find the decomposition of O into O_f and O_b, one must have information on the structure of the synaptic complex DNA which is not bound to the enzyme. This information can come from electron micrographs of the synaptic complex, from gel velocity of the synaptic complex, or from knowledge of the spectrum of structure (supercoiling, knotting, etc.) of the substrate population. For example, from the electron micrograph of figure 11, we can see that O_f = (0) for *resolvase*.

Fortunately, enough experimental information is often available to allow solution of equations (6.4-5). For example, multiple rounds of processive recombination may have been characterized, or the results of a single round of recombination on a variable family of substrates may be known. We make the following mathematical assumption regarding processive recombination:

MATHEMATICAL ASSUMPTION 6.6:   *For n rounds of processive recombination,   post-recombination synaptosome = $(O_b + nR)$.*

That is, for each successive round of processive recombination, we assume that a copy of the recombinant tangle R is added to the synaptosome. Hence, the result of n rounds of processive recombination on the synaptosome $(O_b + P)$ is to replace the parental tangle P by the tangle nR (the sum of n copies of the recombinant tangle R).

Therefore, for processive recombination, our model assumes:

PROCESSIVE ROUND n PRODUCT EQUATION:

(6.7) $N(O_f + O_b + nR) = $ nth round product.

The mathematical assumption of tangle addition for processive recombination allows us to solve the tangle equations, obtaining answers which agree with observations for processive Tn3 resolvase [ES1,S2] and Gin [SE] recombination. Moreover, the assumption that processive recombination can be modelled by tangle addition has important biological implications. If one makes the biologically reasonable assumption that the recombinant tangle R has no local knots, then we have the following:

PRIME PRODUCTS THEOREM 6.8: *The product of n rounds of processive recombination on a single unknotted DNA substrate must be either trivial or a prime knot or link of two components.*

Composite knots and catenanes have been observed in recombination experiments [WD,KK]; the recombination pathway which produces these products must therefore involve distributive recombination on a nontrivial (knotted or catenated) intermediate.

In order to solve the recombination equations (6.4-5), the first step is to prove that the solutions for the variables $\{O_f, O_b, P, R\}$ are rational tangles. Once we know to look for rational solutions, the tangle calculus can be employed to compute the solutions. As we will see below, the analysis of the *Tn3 resolvase* experiments [KS,WC2,WD] makes use of recent results on Dehn surgery on 3-manifolds [CG] to prove that many of the solutions to the tangle equations are rational tangles. On the other hand, if one makes the reasonable biological assumption that the enzyme itself forms a 3-ball, and that the two DNA strands are bound to the surface of the enzyme (before and after recombination), then the tangles $\{O_f, O_b, P, R\}$ will necessarily be rational tangles. This last biological assumption is often unnecessary, since the simplicity of the recombination products forces rationality of many of the tangle solutions [ES,S3].

Another mathematical treatment of site-specific recombination occurs in [WCM], in which the Jones polynomial is used. In [WCM] (in terms of the tangle notation of this paper) it is assumed that the parental tangle $P = (0)$, and that the recombinant tangle $R = (\pm 1)$. Knowledge of the topology of the substrate and product(s) is then used to compute the Jones polynomials of other products. This contrasts with our method, in which we solve tangle equations to obtain information about $\{Of, Ob, P, R\}$.

7. Tn3 Resolvase

Tn3 Resolvase is a site-specific recombinase which reacts with certain circular duplex DNA substrate with directly repeated recombination sites [WC2,WD]. One begins with supercoiled unknotted DNA substrate, and treats it with *resolvase*. Most of the time, *resolvase* mediates one round of recombination, and releases the linked product. The principal product [WC2] of this reaction is the Hopf link of fig. 1E, the 4-plat <2>, and minor products in gel bands corresponding to 4,5,and 6-crossing products were observed. Since we have a substrate with direct repeats, the sites coherently orient the substrate circle, and induce an orientation on each recombination product. By clever manipulation of the experimental protocol, this induced orientation is observable, and it tells us that the Hopf link product has linking number -1 [WC2]. *Resolvase* is known to act **dispersively** in this situation--to bind to the circular DNA, to mediate a single recombination event, and then to release the linked product. It is also known that *resolvase* and free (unbound) DNA links do not react, presumably because the enzyme cannot manage the spatial juxtaposition of the sites when the sites lie on separate molecules. However, in one in 20 encounters, resolvase acts processively, whereby additional recombinant strand exchanges at a single binding encounter are promoted prior to the release of the product, with yield decreasing exponentially with increasing number of recombination rounds. Electron microscopy revealed the minor recombination products to be the 4-crossing figure 8 knot <1,1,2>(figure 12A), the 5-crossing +Whitehead link <1,1,1,1,1>, and the composite 6-crossing link of figure 1H. Given these products, a mechanism for *resolvase* was hypothesized and a new experiment for a predicted 6-crossing knot was performed. The experimental problem was to separate the (relatively rare) 6-crossing knot from the (relatively abundant) 6-crossing link. This was accomplished by 2-dimensional gel electrophoresis, and the discovery of the 6-crossing knot 6_2^* <1,1,1,2,1> substantiated a model for *Tn3 resolvase* mechanism [WD].

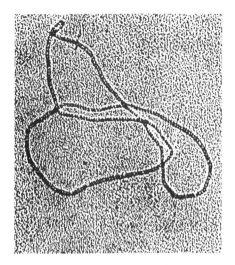 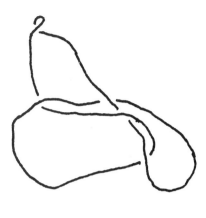

A. Resolvase Figure 8 Knot <1,1,2>
(Courtesy N.R. Cozzarelli and A. Stasiak)

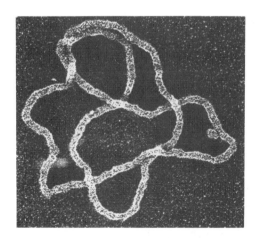 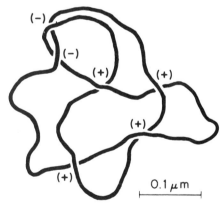

B. Resolvase knot 6_2^* <1,1,1,2,1>
(Courtesy N.R. Cozzarelli)

Figure 12: Resolvase Recombination Products

Starting with recombination products {<2>,<1,1,1,1,1>,<1,1,2>, figure 1H}, our first task is to decide on the recombination pathway for each product. That is, for each product, we must determine how many rounds of recombination and which type of recombination (processive, distributive) is involved. We have direct repeats in our circular substrate, so an odd number of recombination events will produce a pair of circles (a link), and an even number of recombination events will produce a single circle (a knot). We know that the major product (the Hopf link <2>) is the result of one round of recombination; the fact that *resolvase* and free Hopf links do not react means that there is no distributive recombination with the Hopf link as an intermediate; so every recombination pathway begins with at least two rounds of recombination at the first binding encounter. The figure 8 knot must be the result of a even number of recombination events. Since the yield of our recombination reaction is on the order of 5%, the figure 8 knot must be the result of two rounds of processive recombination; the result of 4 rounds of recombination (either processive or distributive) would be present in very small amounts. The +Whitehead link <1,1,1,1,1> is the result of three rounds of recombination (either processive or distributive), since the amount of product of five rounds of recombination will be vanishingly small. The electron micrograph of the *resolvase* synaptic complex in figure 11 shows that $O_f = (0)$, so for resolvase our variables are $\{O_b, P, R\}$. The experimental results of the first two rounds of processive Tn3 recombination yield tangle equations with four solutions for the tangles O_b and R.

THEOREM 7.1 [ES1]: *Suppose that tangles* O_b , *P and R satisfy the following   equations:*
    (i) $N(O_b+P) = <1>$
    (ii) $N(O_b+R) = <2>$
    (iii) $N(O_b+R+R) = <2,1,1>$
Then $\{O_b,R\} = \{(-3,0),(1)\}, \{(3,0),(-1)\}, \{(-2,-3-1),(1)\}$ *or* $\{(2,3,1),(-1)\}$.

Three of the four solutions in theorem 7.1 are mathematically relevant but biologically irrelevant; we would like to find the biologically relevant solution. For this, we absolutely require a chiral product. If all products are achiral (as are <2> and <1,1,2> in 7.1 above), the tangle equations must have an even number of solutions. Fortunately, we have a chiral product, the +Whitehead link <1,1,1,1,1>. If one assumes that the +Whitehead link is the result of three rounds of processive recombination, this can then be used to discard three of these solutions, leaving us with the one believed to be biologically correct (shown in fig. 9), the solution which correctly predicts the result of four rounds of processive recombination (the DNA knot <1,1,1,2,1>). As we expect from theorem 6.8, the composite link of figure 1H cannot arise as a processive recombination product; it is the result of three rounds of distributive recombination, with two rounds of recombination on the unknot at the first binding encounter to produce the figure 8 knot, followed by one round of recombination on the figure 8 knot at the second binding encounter to produce this composite link. Corollary 7.2 below can be viewed as a mathematical proof of *resolvase* synaptic complex structure: the model proposed in [WD] is the only explanation for the first three observed products of Tn3 recombination, assuming that processive recombination acts by adding on copies of the recombinant tangle R.

COROLLARY 7.2 [ES1]: *Suppose that tangles O_b , P and R satisfy the following equations:*

   (i) $N(O_b+P)$ = <1>
   (ii) $N(O_b+R)$ = <2>
   (iii) $N(O_b+R+R)$ = <2,1,1>
   (iv) $N(O_b+R+R+R)$ = <1,1,1,1,1>.
Then $\{O_b,R\}$ = $\{(-3,0),(1)\}$, and $N(O_b+R+R+R+R)$ = <1,1,1,2,1>.

There is another observed chiral product of processive Tn3 recombination. For site-specific recombination on a circular substrate with direct repeats, the orientation of the circular substrate induces an orientation on the DNA link which is the result of one round of recombination, as shown in figure 9. The observed linking number of the DNA Hopf link produced by one round of Tn3 recombination is equal to -1. How can this information be used? Note that there is no information in theorem 7.1 or corollary 7.2 concerning the parental tangle P. Since P appears in only one equation, the substrate equation $N(O_b+ P)$ = <1>, for each rational solution for O_b in this equation, there will be infinitely many rational solutions for P, given by the formula in theorem 4.1. Most biologists believe that P = (0), and a bio-mathematical argument exists for this [SE] . With the additional hypothesis that P = (0), we can determine enzyme mechanism and synaptic complex structure using only two products!

COROLLARY 7.3:   *Suppose that tangles $\{O_b,R\}$ satisfy the following equations:*

   (i) $N(O_b+(0))$ = <1>
   (ii) $N(O_b+R)$ = <2>, with linking number -1
   (iii) $N(O_b+R+R)$ = <2,1,1>
Then $\{O_b,R\}$ = $\{(-3,0),(1)\}$, $N(O_b+R+R+R)$ = <1,1,1,1,1> and $N(O_b+R+R+R+R)$ = <1,1,1,2,1> .

<u>Proof</u>:

From theorem 7.1, we have at most 4 solutions for $\{O_b,R\}$, namely $\{(-3,0),(1)\}$, $\{(3,0),(-1)\}$, $\{(-2,-3-1),(1)\}$, and $\{(2,3,1),(1)\}$. the last two pairs do not satisfy equation (i), because $N((-2,-3,-1)+(0))$ and $N((2,3,1)+(0))$ are a chiral pair of 6-crossing knots. $N((3,0)+(-1))$ is the oriented Hopf link $<2>$ with linking number $+1$, and $N((-3,0)+(1))$ is the oriented Hopf link with linking number -1.

*Proof of theorem 7.1: The first step in the proof is to argue that R must be a rational tangle. Now R, O_b and (O_b+R) are locally unknotted, because $N(O_b+R)$ is the 4-plat $<2>$, in which both components are unknotted. Likewise, P is locally unknotted, because $N(O_b+P)$ is the unknot. Moreover, if B is a prime tangle of parity (0) or (1), and A is any locally unknotted tangle, then $(A+B)$ is likewise a prime tangle [L2]. Now if R has the parity of $(0,0)$, then $N(O_b+R+R)$ must be a link of either 2 or 3 components. Since $N(O_b+R+R)$ is a knot, then R has the parity of (0) or (1). Hence, if R is a prime tangle, then so is (O_b+R). But $N((O_b+R)+R)$ is the 4-plat $<2,1,1>$, so R cannot be a prime tangle (lemma 3.1). This means that R is a rational tangle.

The next step is to argue that O_b is a rational tangle. Suppose that O_b is a prime tangle. Then P must be a rational tangle, because $N(O_b+P)$ is the unknot. The 2-fold branched cyclic cover O_b' is a knot complement. Now R is a rational tangle, so $R+R$ is locally unknotted. Since $N(O_b+(R+R)) = <2,1,1>$, we conclude that $(R+R)$ is a rational tangle. Taking the 2-fold branched cyclic covers, we have that $N(O_b+R)' = L(2,1)$, and $N(O_b+R+R)' = L(5,3)$. Since both R and $R+R$ are rational tangles, this means that Dehn surgery on the knot complement O_b' produces $L(2,1)$ and $L(5,3)$. Application of the cyclic surgery theorem [CG] shows that O_b' must be a solid torus, a contradiction to the assumption that O_b was a prime tangle. We conclude that O_b is a rational tangle.

We will now use the equations (ii) and (iii) to simultaneously compute the rational tangles O_b and R. Since both $<2>$ and $<2,1,1>$ are achiral 4-plats (equal to their mirror images), given any solution $\{O_b,R\}$ for these two equations, then $\{-O_b,-R\}$ is likewise a solution for these equations, where the minus sign indicates the mirror image. If both O_b and R are non-integral, then $N(O_b+R+R)$ is a nontrivial Montesinos knot, and cannot be a 4-plat. Since $N(O_b+R+R) = <2,1,1>$ is a 4-plat, we conclude that exactly one of $\{O_b,R\}$ must be an integral tangle, and that $O_b \neq (0,0) \neq R$.

Suppose that R is the integral tangle (r), so $(R+R) = (2r)$. If $r = 0$, then $<2> = N(O_b+(0)) = N(O_b+(0)+(0)) = <2,1,1>$, a contradiction. If $R = r$ and $O_b = u/v$ is a solution for equations (ii) and (iii), then the integers u,v and r satisfy the following two equalities:

$$\left| u + rv \right| = 2 \quad \text{and} \quad \left| u + 2rv \right| = 5 \ .$$

These equations have the following solutions: $\{(u,rv)\} = \{(-1,3)\}$, $\{(1,-3)\}$, $\{(9,-7)\}$, and $\{(-9,7)\}$. From the first solution set, $u = -1$ and $rv = 3$, so as possible values for solutions we have $\{(u/v,r)\} = \{(-1/3,1)\}$, $\{(1/3,-1)\}$, $\{(-1,3)\}$ and $\{(1,-3)\}$. The last two are not solutions to (iii), and are discarded. Working through all the values for $\{(u,rv)\}$ and discarding extraneous solutions, we find that we are left with the solutions stated in the theorem. This means that $\{(-3,0),(1)\}$, $\{(3,0),(-1)\}$, $\{(-2,-3,-1),(1)\}$ and $\{(2,3,1),(-1)\}$ form a complete set of solutions for $\{O_b,R\}$ in equations (ii) and (iii), if R is integral.

Suppose now that O_b is integral, $O_b = (s)$ and $R = u/v$ ($v > 1$) form a set of solutions to $\{(ii),(iii)\}$. We have $| u + sv | = 2$. Since O_b is integral, then $N((O_b+R)+R) = N(R+(R+O_b)) = <2,1,1>$. The rational tangle $(R+O_b)$ is classified by the rational number $(s + u/v)$, so $| 2uv + sv^2 | = v| 2u + sv | = 5$, so $v = 5$. Simultaneous solution of the equations $\{ | u + 5s | = 2, | 2u + 5s | = 1 \}$ yields the following solutions: $\{(u,s)\} = \{(-1,3/5), (1,-3/5), (-3,1), (3,-1)\}$. The first two solutions are discarded because s is not an integer; the last two solutions are discarded because they do not yield solutions to equation (iii).

REFERENCES

[A] K. Abremski, B. Frommer and R.H. Hoess. Linking-number changes in the DNA substrate during Cre-mediated loxP site-specific recombination. J. Mol. Biol. **192**(1986), 17-26.

[Ar] T. Archibald. Connectivity and smoke-rings:Green's second identity in its first fifty years. Mathematics Magazine **62**(1989), 219-232.

[BCW] W.R. Bauer, F.H.C. Crick and J.H. White. Supercoiled DNA. Scientific American **243**(1980), 100-113.

[BM] H.W. Benjamin, M.M. Matzuk, M.A. Krasnow and N.R. Cozzarelli. Recombination site selection by Tn3 resolvase: topological tests of a tracking mechanism. Cell **40**(1985), 147-158.

[BC] H.W. Benjamin and N.R. Cozzarelli. Geometric arrangements of Tn3 resolvase sites. J. Biol. Chem. **265**(1990), 6441-6447.

[BS] F. Bonahon and L.C. Siebenmann. New Geometric Splittings of Classical Knots. Lon. Math. Soc. Monographs (to appear).

[BZ] G. Burde and H. Zieschang. Knots. de Gruyter (1985).

[C] J.H. Conway. On enumeration of knots and links and some of their related properties. In Computational Problems in Abstract Algebra; Proc. Conf. Oxford 1967. Pergamon (1970), 329-358.

[CK] N.R. Cozzarelli, M.A. Krasnow, S.P. Gerrard and J.H. White. A topological treatment of recombination and topoisomerases. Cold Spring Harbor Symp. Quant. Biol. **49**(1984), 383-400.

[CF] R.H. Crowell and R.H. Fox. Introduction to Knot Theory. Graduate Texts in Mathematics **57**, Springer-Verlag (1977).

[CG] M.C. Culler, C.M. Gordon, J. Luecke and P.B. Shalen. Dehn surgery on knots. Ann. of Math. **125**(1987), 237-300.

[CM] R. Craigie and K. Mizuuchi. Role of DNA topology in Mu transposition: mechanism of sensing the relative orientation of two DNA segments. Cell **45**(1986), 793-800.

[DS] F.B. Dean, A. Stasiak, T. Koller and N.R. Cozzarelli. Duplex DNA knots produced by Escherichia coli topoisomerase I. J. Biol. Chem. **260**(1985), 4795-4983.

[DC] P. Droge and N.R. Cozzarelli. Recombination of knotted substrates by Tn3 resolvase. Proc. N.A.S. USA **86**(1989), 6062-6066.

[EH] P.T. Englund, S.L. Hajduk and J.C. Marini. The molecular biology of trypanosomes. Ann. Rev. Biochem. (1982), 695-726.

[ES] C. Ernst and D.W. Sumners. The growth of the number of prime knots. Math. Proc. Camb. Phil. Soc. **102** (1987), 303-315.

[ES1] C. Ernst and D.W. Sumners. A calculus for rational tangles: applications to DNA recombination. Math. Proc. Camb. Phil. Soc. **108** (1990), 489-515.

[GeN] M. Gellert and H. Nash. Communication between segments of DNA during site-specific recombination. Nature **325**(1987), 401-404.

[GrN] J.D. Griffith and H.A. Nash. Genetic rearrangement of DNA induces knots with a unique topology: implications for the mechanism of synapsis and crossing-over. Proc. N.A.S. USA **82**(1985), 3124-3128.

[HJ] K.A. Heichman and R.C. Johnson. The Hin invertasome: protein-mediated joining of distant recombination sites at the enhancer. Science **249**(1990), 511-517.

[KK] R. Kanaar, A. Klippel, E. Shekhtman, J.M. Dungan, R. Kahmann and N.R. Cozzarelli. Processive recombination by the phage Mu gin system: implications for mechanisms of DNA exchange, DNA site alignment, and enhancer action. Cell **62**(1990), 353-366.

[KP] R. Kanaar, P. van de Putte and N.R. Cozzarelli. Gin-mediated DNA inversion: product structure and the mechanism of strand exchange. Proc. N.A.S. USA **85**(1988), 752-756.

[K] L.H. Kauffman. Knots and physics. This volume.

[K1] L.H. Kauffman. On Knots. Princeton Univ. Press (1987).

[Kn] C.G. Knott. Life and scientific work of P.G. Tait. Cambridge University Press (1911).

[KS] M.A. Krasnow, A. Stasiak, S.J. Spengler, F.Dean, T. Koller and N.R. Cozzarelli. Determination of the absolute handedness of knots and catenanes of DNA. Nature **304**(1983), 559-560.

[LS1] J. Langer and D.A. Singer. Curve straightening and a minimax argument for closed elastic curves. Topology **24**(1985), 75-88.

[LS2]   J. Langer and D.A. Singer. Knotted elastic curves in R^3. J. London Math. Soc. **30**(1984), 512-520.

[L1]   W.B.R. Lickorish. Polynomials for links. Bull. L.M.S. **20**(1988), 558-588.

[L2]   W.B.R. Lickorish. Prime knots and tangles. Trans. A.M.S. **267**(1981), 321-332.

[MM]   J.C. Marini, K.G. Miller and P.T. Englund. Decatenation of kinetoplast DNA by topoisomerases. J. Biol. Chem. **255**(1980), 4976-4979.

[P]   W.F. Pohl, DNA and differential geometry. The Mathematical Intelligencer **3**(1980), 20-27.

[MC]   K. Mizuuchi and R. Craigie. Mechanism of bacteriophage Mu transposition. Ann. Rev. Genet. (1986), 385-429.

[PN]   T.J. Pollock and H.A. Nash. Knotting of DNA caused by genetic rearrangement: evidence for a nucleosome-like structure in site-specific recombination of bacteriophage lambda. J. Mol. Biol. **170**(1983), 1-18.

[RE]   C.A. Rauch, P.T. Englund, S.J. Spengler, N.R. Cozzarelli and J.H. White. Kinetoplast DNA: structure and replication (in preparation).

[R]   D. Rolfsen. Knots and Links. Publish or Perish, Inc. (1990).

[SD]   D. Sherratt, P. Dyson, M. Boocock, L. Brown, D. Summers, G. Stewart and P. Chan. Site-specific recombination in transposition and plasmid stability. Cold Spring Harbor Symp. Quant. Biol. **49**(1984), 227-233.

[SK]   K. Shishido, N. Komiyama and S. Ikawa. Increased production of a knotted form of plasmid pBR322 DNA in Escherichia coli DNA topoisomerase mutants. J. Mol. Biol. **195**(1987), 215-218.

[SSC]   S.J. Spengler, A. Stasiak and N.R. Cozzarelli. The stereostructure of knots and catenanes produced by phage lambda integrative recombination: implications for mechanism and DNA structure. Cell **42**(1985), 325-334.

[SSS]   S.J. Spengler, A. Stasiak, A.Z. Stasiak and N.R. Cozzarelli. Quantitative analysis of the contributions of enzyme and DNA to the structure of lambda integrative recombinants. Cold Spring Harbor Symp. Quant. Biol. **49**(1984), 745-749.

[SSB]   W.M. Stark, D.J. Sherratt and M.R. Boocock. Site-specific recombination by Tn3 resolvase: topological changes in the forward and reverse reactions. Cell **58**(1989), 779-790.

[S1]   D.W. Sumners. Knots, macromolecules and chemical dynamics. In Graph Theory and Topology in Chemistry, Elsevier (1987), 297-318.

[S2]   D.W. Sumners. The role of knot theory in DNA research. In Geometry and Topology, Marcel Dekker (1987), 297-318.

[S3]   D.W. Sumners. Untangling DNA. The Mathematical Intelligencer **12**(1990), 71-80.

[SE]   D.W. Sumners, C.E. Ernst, N.R. Cozzarelli and S.J. Spengler. The tangle model for enzyme mechanism. (in preparation).

[T]   W. Thompson. On vortex atoms. Philosophical magazine **34**(July, 1867), 15-24.

[W]   D.M. Walba. Topological Stereochemistry. Tetrahedron **41**(1985), 3161-3212.

[Wa1]   J.C. Wang. DNA topoisomerases. Ann. Rev. Biochem. (1985), 665-697.

[Wa2]   J.C. Wang. DNA topoisomerases. Scientific American **247**(1982), 94-109.

[WC1]   S.A. Wasserman and N.R. Cozzarelli. Biochemical topology: applications to DNA recombination and replication. Science **232**(1986), 951-960.

[WC2]   S.A. Wasserman and N.R. Cozzarelli. Determination of the stereostructure of the product of Tn3 resolvase by a general method. Proc. N.A.S.(USA) **82**(1985), 1079-1083.

[WD]   S.A. Wasserman, J.M. Dungan and N.R. Cozzarelli. Discovery of a predicted DNA knot substantiates a model for site-specific recombination. Science **229**(1985), 171-174.

[Wh]   J.H. White. An introduction to the geometry and topology of DNA structure. In Mathematical Methods for DNA Sequences, CRC Press (1989), 225-253.

[WC]   J.H. White and N.R. Cozzarelli. A simple topological method for describing stereoisomers of DNA catenanes and knots. Proc. N.A.S. USA **81**(1984), 3322-3326.

[WM]   J.H. White, K.C. Millett and N.R. Cozzarelli. Description of the topological entanglement of DNA catenanes and knots by a powerful method involving strand passage and recombination. J. Mol. Biol. **197** (1987), 585-603.

Department of Mathematics, Florida State University, Tallahassee, FL 32306-3027

E-mail address: sumners@math.fsu.edu

Proceedings of Symposia in Applied Mathematics
Volume 45, 1992

Topology of Polymers

Stuart G. Whittington

Abstract

Polymer molecules can be thought of as long flexible strings
which can be highly self-entangled. There are interesting questions about the entanglement complexity both when the polymer
can be modelled as a simple closed curve, and when it should be
modelled by a graph with some other homeomorphism type. A
convenient model of such polymers is a graph of some fixed homeomorphism type (for instance the circle graph), embedded in a
three dimensional lattice (such as Z^3). An embedding of the circle
graph is a *polygon*, and we can write p_n for the number of polygons
with n edges in Z^3, where two polygons are considered distinct if
they cannot be superimposed by translation. If p_n^o is the number
of n-edge polygons which are *unknotted*, how does the ratio p_n^o/p_n
behave? It can be shown that this ratio goes to zero exponentially
rapidly, as $n \to \infty$. This paper will examine the methods used in
this proof, and their extensions to other homeomorphism types. In
addition, one can ask how badly knotted a typical polygon will be.
Again some rigorous results are available.

For more complicated graphs (such as the θ-graph) one can
ask about other knotted structures. We call an embedding *almost
unknotted* if the embedding is knotted but if it becomes unknotted
upon deleting any edge. Some results will be presented on the
numbers of such embeddings, again in the $n \to \infty$ limit.

None of these approaches say much about the behaviour for
moderate values of n. To address this question, the only technique
currently available seems to be Monte Carlo methods. One such
approach will be outlined, and some results reviewed.

1991 *Mathematics Subject Classification*. Primary 57M25; Secondary 82B41, 82D60.

This paper is in final form and no version of it will be submitted for publication
elsewhere.

1   Introduction

Self-avoiding walks on regular lattices have been studied for many years as a model of linear polymer molecules in dilute solution. They have a high degree of conformational freedom, and the self-avoiding condition mimics the repulsive interaction between pairs of monomers in the polymer. Some results about self-avoiding walks will be reviewed in Section 2. In a similar way, ring polymers can be modelled as polygons embedded in lattices, and polymers with more exotic architectures can be modelled by graphs having other homeomorphism types. Section 2 will also contain some background results on the numbers of embeddings of these graphs.

As linear polymers move in solution they can become self-entangled. If the polymers undergo a ring closure reaction these entanglements can be captured as knots in the resulting ring polymers and the original entanglements in the linear case can be investigated by studying the knots in the resulting rings. Why are polymer physicists interested in entanglement? One reason is their interest in making highly crystalline polymers. One way to do this is to crystallize polymers from solution. If the polymer is self-entangled, this entanglement may be preserved in the crystallization process producing a defect or fault in the crystal. In fact it is thought that such entanglements are concentrated in the amorphous regions of the polymer crystal. In addition, topological entanglements within or between polymer chains may contribute to rheological properties of the polymer.

The phenomenon of knotting in ring polymers is not well understood. Frisch and Wasserman [1] and Delbruck [2] conjectured that sufficiently long ring polymers would be knotted with probability one, and since then the problem has been studied numerically by several groups [3, 4]. It was not until 1988 that the validity of the Frisch-Wasserman-Delbruck conjecture was established rigorously, for a lattice model of a polymer, by Sumners and Whittington [5] and, independently, by Pippenger [6]. For a self-avoiding polygon with n edges on the simple cubic lattice, these papers establish that the probability $P(n)$ that the polygon is knotted goes to unity exponentially rapidly as n goes to infinity. One can also ask about the complexity of the knot. For instance, is the knot likely to be prime or composite? How does a suitable measure of

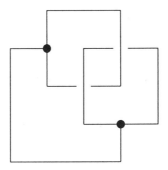

Figure 1: A dumbell graph with linked cycles

the entanglement complexity, such as the average value of the crossing number or unknotting number, depend on n? These questions have been addressed by Soteros, Sumners and Whittington [7]. Some of this work will be discussed in Section 3.

Polymers can have more complicated architectures than a simple linear or ring polymer, and in these cases should be modelled by more complicated types of graph. One approach is to model the polymer by a graph of some fixed homeomorphism type, with a total of n edges, embedded in Z^3. One can then ask similar questions about the probability that the embedding is knotted, and about the entanglement complexity. These questions have also been examined in [7]. The approach used was to focus on circuits in the graph, and to show that at least one such circuit would contain a knot with high probability (see Section 3). We call such knots *local* since they are associated with a particular circuit. In certain circumstances, embeddings of graphs can be knotted even though they contain no local knots [8, 9]. See for instance the embedding of the *dumbell* or *handcuff* graph shown in Figure 1. Kinoshita has introduced the term *almost unknotted* graphs for embeddings which are knotted, but which become unknotted upon deleting any edge. One can ask how likely it is that an embedding will be almost unknotted, given that it contains no local knots. We describe some results on this question in Section 4.

The results which will be discussed in Sections 3 and 4 are all asymptotic in nature. I.e. they address the behaviour in the limit $n \to \infty$. They say little about knot probabilities for small values of n. To address this question, the only tools currently available seem to be Monte Carlo approaches. Various Monte Carlo methods have been used by a number of groups but all of them have to cope with the difficulty of generating a uniformly distributed random sample of (for instance) polygons on a lattice. Very recently Madras et al [10] and Janse van Rensburg et al [11] have developed a version of the pivot algorithm for self-avoiding walks [12, 13], which samples along a realization of a Markov chain defined on the space of polygons. This algorithm has been used [14] to investigate the n-dependence of the knot probability for polygons on the face-centred cubic lattice and, in particular, to estimate the value of the constant appearing in the exponential approach term. This work is described in Section 5.

In Section 6 we discuss a problem which, up to now, has only been approached using Monte Carlo methods. This is the question of the dependence of the dimensions of a polygon on its knot type. It might be a question of some practical importance since circular DNA molecules with different knot types are isolated using methods which rely on size differences.

Finally, the problem of how one might characterise the complexity of the self-entanglements in a self-avoiding walk is discussed in Section 7.

For a very readable and modern account of knot theory and its applications see [15]. Other excellent sources for results in knot theory include the books by Rolfsen [16] and by Burde and Zieschang [17]. Background definitions and results in graph theory can be found in Ore's book [18]. An excellent treatment of Monte Carlo methods is available in the book by Hammersley and Handscomb [19].

2   Self-avoiding walks and related structures

In this section we review some background results about self-avoiding walks, and their relation to polygons and other graphs embeddable in lattices. We shall be chiefly concerned with Z^3, the simple cubic lattice. The vertices of this lattice are the integer points in R^3, and the edges join pairs of vertices which are unit distance apart. A *walk* is a directed sequence of edges, such that adjacent pairs of edges in the sequence are incident on a common vertex. A walk is *self-avoiding* if no vertex is visited more than once. We shall count walks up to translation. That is, two walks will be considered distinct if they cannot be superimposed (with due regard for the directed nature) by translation. If we write c_n for the number of self-avoiding walks with n edges on Z^3, it is easy to see that $c_1 = 6$, $c_2 = 30$, $c_3 = 150$, $c_4 = 5 \times c_3 - 24 = 726$, etc. Rather little is known about the asymptotic behaviour of c_n, but Hammersley [20] has proved the following important theorem:

Theorem 2.1

$$0 < \lim_{n \to \infty} n^{-1} \log c_n = \inf_{n > 0} n^{-1} \log c_n \equiv \kappa < \infty \qquad (2.1)$$

where κ is called the connective constant of the lattice.

Proof: The idea behind this proof is both simple and powerful. Consider a self-avoiding walk of m steps and a self-avoiding walk of n steps. Concatenate these so that the last vertex of the first walk coincides with the first vertex of the second walk. The first walk can be chosen in c_m ways and the second in c_n ways, and the resulting graphs include all self-avoiding walks of $m + n$ steps, so that

$$c_m c_n \geq c_{m+n}. \qquad (2.2)$$

Since every walk consisting only of steps in the *positive* coordinate directions is self-avoiding, $c_n \geq 3^n$. This, together with (2.2), implies (2.1).

This result is important in that it establishes that the number of walks increases exponentially rapidly. However, it says nothing about the rate of approach to the limit in (2.1). This is an interesting question in its own right. In fact, there are strong reasons to believe that

$$c_n = \exp(\kappa n + O(\log n)). \qquad (2.3)$$

This is supported by exact enumeration and Monte Carlo results [21], as well as by a formal connection with the theory of critical phenomena [22, 23]. A proof of this result would be a major advance in the theory of self-avoiding walks. Slade [24] has shown that in sufficiently high dimension

$$c_n = A e^{\kappa n} (1 + o(1)) \qquad (2.4)$$

and Hara and Slade [25] have recently established this result in five and higher dimensions. There is also a recent contribution by Hammersley [26] to the behaviour in lower dimensions.

Since we are interested in the properties of ring polymers the corresponding model is a self-avoiding polygon. We write p_n for the number of (undirected, unrooted) self-avoiding polygons embeddable in a lattice. Again we count two polygons as distinct if they cannot be superimposed by translation. For instance, on the simple cubic lattice, $p_4 = 3$, $p_6 = 22$ and $p_8 = 207$. Hammersley [27] has shown that

$$\lim_{n \to \infty} n^{-1} \log p_n = \kappa \tag{2.5}$$

where κ is the connective constant defined in equation (2.1).

We shall also be interested in more exotic graphs, in particular θ graphs (so called because of their resemblance to the Greek letter), and generalised θ graphs. Consider a graph with a fixed homeomorphism type which we denote by τ. Let $g(n, \tau)$ be the number of embeddings in Z^3 (up to translation) of the graphs homeomorphic to τ, with a total of n edges. For instance $g(7, \theta) = 18$ and $g(8, \theta) = 24$. Provided that τ has no vertex of degree greater than six [7]

$$\lim_{n \to \infty} n^{-1} \log g(n, \tau) = \kappa. \tag{2.6}$$

3   Rigorous results on knot probabilities

In this section we first discuss a rigorous result which says that almost all sufficiently long polygons are knotted. The idea is conceptually very simple, and falls into three parts. The first is a pattern theorem due to Kesten [28]. A *pattern* is any finite self-avoiding walk. Given a particular pattern γ, if there exists a self-avoiding walk on which the pattern γ appears three times then we call γ a *K-pattern*. (The point of appearing three times is that one of the occurences must be "between" the other two occurences, and so there must be a way in to the beginning of the pattern and a way out from the end. Contrast a tight spiral which can occur at the beginning or end of a walk, but nowhere else.) Roughly speaking, Kesten's theorem says that every K-pattern will appear with positive density on almost all sufficiently long self-avoiding walks.

Theorem 3.1 *For any K-pattern γ, let $c_n(\epsilon; \gamma)$ be the number of n-step self-avoiding walks on which γ appears at most ϵn times. Then there exists a value of $\epsilon > 0$ for which $\limsup_{n \to \infty} n^{-1} \log c_n(\epsilon; \gamma) < \kappa$.*

Proof: The proof can be found in [28].

Since we shall be primarily concerned with polygons we need a corresponding theorem which applies to that case. Let γ be an undirected pattern. We now say that γ is a K-pattern if there exists a self-avoiding walk on which one of the two directed versions of the pattern appears three times.

Theorem 3.2 *Let $p_n(\epsilon; \gamma)$ be the number of n-edge polygons on which the K-pattern γ appears at most ϵn times. Then there exists some $\epsilon > 0$ for which*

$$\limsup_{n \to \infty} n^{-1} \log p_n(\epsilon; \gamma) < \kappa.$$

Proof: Define the *bottom vertex* of a polygon by lexicographic ordering. The vertex has two incident edges, each incident on a second vertex. Choose the lower of these two vertices (lexicographically) and delete the corresponding edge. This gives an (undirected) self-avoiding walk with $n-1$ edges. Since deleting an edge cannot create a pattern, the theorem follows from Theorem (3.1) and equation (2.5).

The second ingredient in the proof is the idea of a knotted arc. We can capture the important part of a knot such as a trefoil and tie it so tightly on the lattice that the rest of the walk cannot pass through the neighbourhood of this subwalk and untie the knot. Technically, the subwalk and its associated dual 3-cells form a *knotted ball pair*, i.e. a ball pair (B^3, B^1) which is not ambient isotopic to the standard ball pair. These knotted ball pairs can be constructed so that the subwalk (B^1) is a K-pattern, leading to the following Lemma:

Lemma 3.1 *Every knot type is represented by a K-pattern on Z^3 such that the K-pattern and its associated dual 3-cells form a knotted ball pair.*

Finally we need a standard result [29] in knot theory:

Lemma 3.2 *For a given non-trivial knot k there is no knot l such that the connected sum $k \# l$ is unknotted.*

We are now ready to state the primary result of this section as a theorem:

Theorem 3.3 *The probability $P(n)$ that an n-gon is knotted is given by*

$$P(n) = 1 - \exp(-\alpha n + o(n)) \tag{3.1}$$

where $\alpha > 0$.

Proof: If we write p_n^o for the number of *unknotted* polygons, a concatenation argument shows that the limit $\lim_{n \to \infty} n^{-1} \log p_n^o$ exists so that

$$p_n^o = e^{\kappa_o n + o(n)}. \tag{3.2}$$

The set of unknotted polygons with n edges is a subset of the polygons which do not contain a trefoil, which, in turn, are a subset of the polygons which do not contain the K-pattern associated with a trefoil. Hence $\kappa_o < \kappa$. Since $P(n) = 1 - p_n^o/p_n$, this gives eqn (3.1) with $\alpha = \kappa - \kappa_o > 0$, which proves the theorem.

Since every knot type has an associated K-pattern which, with its associated dual 3-cells, forms a knotted ball pair, every knot type must appear with positive density on almost all sufficiently long polygons. In particular, almost all polygons contain highly composite knots. Not only are unknotted polygons exponentially rare, but so are polygons with only prime knots.

This immediately allows us to say something about the numbers of polygons with *fixed* knot type. Suppose that $p_n(3_1)$ is the number of n-gons which are trefoils. What can we say about the asymptotic behaviour for large n? The above argument implies that

$$\limsup_{n \to \infty} n^{-1} \log p_n(3_1) < \kappa. \tag{3.3}$$

By concatenating a fixed trefoil with an unknotted polygon, it is easy to see that

$$\liminf_{n \to \infty} n^{-1} \log p_n(3_1) \geq \kappa_o \tag{3.4}$$

but the existence of the limit has not been established. If k is a (composite) knot which has l as a factor, similar arguments can be used to show that

$$\limsup_{n \to \infty} n^{-1} \log p_n(l) \leq \liminf_{n \to \infty} n^{-1} \log p_n(k). \tag{3.5}$$

However, there are no results on the relative numbers of polygons with different prime knots.

One can seek to characterise the average complexity of the knotted polygons. It turns out [7] that any reasonable measure of entanglement complexity (e.g. the average value of the crossing number, unknotting number, span of any knot polynomial, etc.) goes to infinity as $n \to \infty$. In fact one can show that these quantities diverge at least linearly in n. This is essentially because these numbers are additive for trefoils and Theorem 3.2 and Lemma 3.1 show that n-gons contain at least ϵn trefoils, for some postive ϵ.

Similar but slightly weaker results have been proved for Gaussian random polygons in R^3 [30].

It is not too difficult [7] to extend these results to embeddings in Z^3 of certain types of graphs, other than the circle graph. Consider a planar graph of fixed homeomorphism type τ, having no cut edges, and having no vertex of degree greater than 6. Since no vertex is of degree greater than 6, the graph has an embedding in Z^3. We next need to decide on what we mean by an unknotted embedding. Since the graph is planar, there is no difficulty since all planar embeddings of the graph in R^3 are equivalent, and we call an embedding unknotted if it is a member of this equivalence class.

Theorem 3.4 *For sufficiently large n, all but exponentially few embeddings of τ in Z^3, with a total of n edges, are knotted.*

Proof: Since the graph has no cut edges every edge must be in a cycle. Since the homeomorphism type is fixed, so is the cyclomatic index (or first Betti number). Hence there must be a largest cycle which contains a non-zero fraction of the edges. By the arguments given above this cycle is almost always knotted and so the embedding of the graph is almost always non-planar, and hence almost always knotted.

In fact the restriction to planar graphs is unnecessary. Some graphs are intrinsically complex in terms of their embeddings in 3-space. For instance [31] every embedding of K_6 (the complete graph on six vertices) contains a pair of homologically linked cycles. We define $N_0(\tau)$ to be the minimum crossing number over all embeddings of the graph τ in R^3, and define an equivalence class of PL embeddings of representatives of τ to be *unknotted* if there are members of that equivalence class which realise $N_0(\tau)$. With this definition of unknotted, Theorem 3.4 is still valid.

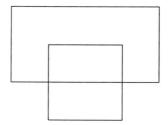

Figure 2: A planar embedding of θ_4

4   Almost unknotted embeddings

Kinoshita [8, 9] has pointed out that it is possible for an embedding of a graph (with cyclomatic index at least equal to two) to be knotted even though none of its cycles contain a knot. Perhaps the simplest example to think of is the dumbell or handcuff graph which has two cycles joined by an edge. This graph can be embedded in R^3 (or in Z^3) so that the two cycles are linked (see Fig. 1.)

Kinoshita was concerned with a more subtle case which he called *almost unknotted* embeddings. An embedding is almost unknotted if it is knotted but becomes unknotted upon deleting any edge. Diagrams for several cases can be found in [32] and Kawauchi has shown that such embeddings exist for any planar graph with no cut edges [33].

We shall be concerned here only with the special case of θ_k graphs, i.e. graphs with two vertices of degree k, joined by k edges. In particular we shall ask for the number of embeddings in Z^3 of graphs homeomorphic to θ_k with some fixed number (n) of edges. Figure 2 shows a planar embedding of θ_4. If we write $\theta_k(n)$ for the number of such embeddings (modulo translation) then $\theta_3(7) = 18$, $\theta_4(10) = 12$, etc.

From (2.6) we have

$$\lim_{n \to \infty} n^{-1} \log \theta_k(n) = \kappa \qquad (4.1)$$

provided that $k \leq 6$. We now ask for the number, $\theta_4^{oo}(n)$, of embeddings of θ_4 which are unknotted (i.e. ambient isotopic to the planar embedding shown in Figure 2), and the number, $\theta_4^o(n)$ of embeddings in which none of the cycles are knotted.

Lemma 4.1

$$\lim_{n \to \infty} n^{-1} \log \theta_4^{oo}(n) = \lim_{n \to \infty} n^{-1} \log \theta_4^o(n) = \kappa_o. \qquad (4.2)$$

Proof: Clearly $\theta_4^{oo}(n) \leq \theta_4^o(n)$. To get a lower bound on $\theta_4^{oo}(n)$ we concatenate a planar embedding of θ_4 with m edges and an unknotted polygon with $n - m$ edges (for instance the embedding shown in Figure 2, rotated into the appropriate plane). We translate so that the left-most plane of the polygon is one lattice space to the right of the right-most plane of the planar θ_4, then add and delete edges to give a new θ_4-graph with n edges, ambient isotopic to the planar embedding. This gives

$$p_{n-m}^o \leq \theta_4^{oo}(n). \tag{4.3}$$

To get an upper bound on $\theta_4^o(n)$ we root an unknotted polygon with m edges and an unknotted polygon with n edges, and translate so that these two roots become coincident. The resulting graphs have two cycles which are unknotted and, if we sum over m, include all the unknotted θ_4's. Hence

$$\theta_4^o(n) \leq \sum_m m(n - m)p_m^o p_{n-m}^o. \tag{4.4}$$

It only remains to take logarithms in (4.3) and (4.4), divide by n and let n go to infinity.

We write $\theta_4^*(n)$ for the number of embeddings of θ_4 (with n edges) which are almost unknotted. Then we have the following lemma:

Lemma 4.2

$$\lim_{n \to \infty} n^{-1} \log \theta_4^*(n) = \kappa_o. \tag{4.5}$$

Proof: Clearly $\theta_4^*(n) \leq \theta_4^o(n)$. To get a lower bound we concatenate an unknotted polygon with $n - m$ edges with an almost unknotted embedding of θ_4 with m edges, similar to that shown in Figure 3. (The figure is drawn to indicate how it can be embedded in Z^3.) This gives

$$\theta_4^*(n) \geq p_{n-m}^o \tag{4.6}$$

and these two bounds are enough to prove the theorem.

This is an interesting and perhaps surprising result. Although we know from Section 3 that *unknotted* embeddings of graphs are exponentially rare compared to *knotted* embeddings, the results in this section show that (at least for θ_4), *almost unknotted* embeddings and *unknotted* embeddings grow at the same exponential rate. These results can be extended in two ways [34]. First, similar theorems can be proved for

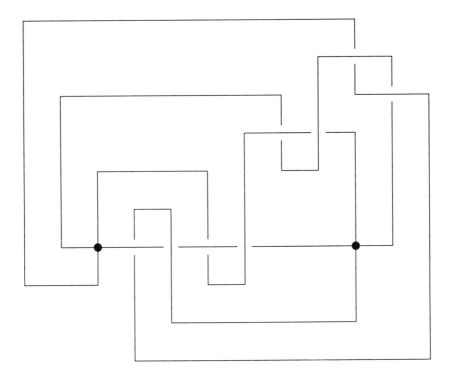

Figure 3: An almost unknotted embedding of θ_4

any homeomorphism type provided that (i) the graph has no cut edges, (ii) it is Eulerian and (iii) (so that it can be embedded in Z^3) it has no vertices of degree greater than six.

A second interesting extension is to the case of *uniform* embeddings, in which the number of edges in each *branch* is equal. If we write $\Theta_k(n)$ for the number of uniform embeddings of the graph θ_k in Z^3 with a *total* of n edges, then clearly $\Theta_k(n) = 0$ unless k divides n, $\Theta_3(9) = 20$, $\Theta_4(12) = 3$, etc. Soteros [35] has shown that

$$\lim_{n \to \infty} n^{-1} \log \Theta_k(n) = \kappa \tag{4.7}$$

for $k \le 6$.

Similar theorems work in this case. However, instead of concatenating a single unknotted polygon and a fixed (unknotted or almost unknotted) embedding of, for instance, θ_4, one needs to concatenate polygons onto each branch. Clearly these polygons must not link one another and this can be achieved by confining them to disjoint wedges, and making use of a theorem [36] about the numbers of polygons in these wedges.

5   Monte Carlo results

The difficulty with the results described in the previous sections is that they do not tell us how likely a polygon is to be knotted when the number of edges in the polygon is small. Or, putting it another way, how large does the polygon have to be before it is knotted with probability 0.1 or 0.5? Currently there seems to be no way to answer this question without resorting to numerical approaches such as Monte Carlo methods.

The idea of a Monte Carlo approach is to generate a random sample of polygons, and determine what fraction of the sample is knotted. The difficulty is to generate a sample which is really random in a reasonable amount of computer time. Early work [3] established that the knot probability in lattice polygons was quite small when the number of edges in the polygon was of the order of a few hundred.

Recently Madras and Sokal [13] showed that an algorithm (now known as the pivot algorithm) which was invented by Lal [12] is extremely efficient at generating a random (but correlated) sample of self-avoiding walks. This algorithm is based on the Metropolis sampling method, in which a Markov chain is defined on the set of self-avoiding walks with a fixed number of edges. The Markov chain must be irreducible (so that every walk can be reached in the sampling) and reversible (so that every walk has the same probability of appearing in the sample). One then produces a realization of the Markov chain, and the states visited in this realization are a (correlated) random sample of the self-avoiding walks. The algorithm can be adapted [10, 11] to work for the polygon problem and has been used to sample polygons on the face centred cubic lattice and to determine the knot type of the polygons in the sample [14].

For moderate values of n the knot probability for polygons on the face-centred cubic lattice turns out to be quite small. E.g. for $n = 800$, $P(n)$ is only about 4×10^{-3}, and this value rises to about 1.2×10^{-2} for $n = 1600$. Even at this value of n almost all the knots found in the Monte Carlo calculations are trefoils (although we know from Section 3 that in the large n limit almost all knots must be composite). The data are well described by the relation

$$1 - P(n) = Ce^{-\alpha n} \tag{5.1}$$

where $\alpha = (7.6 \pm 0.9) \times 10^{-6}$ and C is about 1.

A similar calculation has been carried out for polygons on the simple cubic lattice. In this case $P(800)$ is about 2.8×10^{-3} and $P(1600)$ is about 7×10^{-3}. At $n = 1600$ more that 95% of the non-trivial knots in the sample are trefoils. The data are again well represented by the above equation and α is estimated to be $\alpha = (5.7 \pm 0.5) \times 10^{-6}$ with C being once again close to 1. Unfortunately the results do not resolve the question of the possible lattice dependence of α. All that can be said is that the 95% confidence intervals have substantial overlap.

These results (for pure self-avoiding polygons) model polymers in *good solvents*, i.e. in the situation in which the repulsive interactions between pairs of monomers in the polymer dominate all other interactions. (The situation is in reality more complicated and one should take into account monomer-monomer, monomer-solvent and solvent-solvent interactions. At high temperatures these (sometimes competing) interactions can be represented by a repulsive pseudo-potential between pairs of monomers.) By changing the solvent (or lowering the temperature) one can cause the polymer to undergo a transition to a compact state, in which the number of monomer-solvent interactions is reduced, and the number of monomer-monomer interactions increases. This situation is refered to as *poor solvent* conditions. One can mimic the effect of decreasing the solvent quality by incorporating an attractive potential between neighbouring pairs of vertices of the polygon, i.e. by including a contact potential in the model. If the energy per contact is ϵ we write

$$\phi = -\epsilon/k_B T \qquad (5.2)$$

where k_B is Boltzmann's constant and T is the absolute temperature. Good solvent conditions correspond to $\phi = 0$ and increasing values of ϕ simulate an increasingly bad solvent. The polygons are not equally weighted but, if a polygon has m pairs of vertices which are unit distance apart on the lattice, it has a weight proportional to $e^{m\phi}$.

As ϕ increases the knot probability increases quite dramatically. E.g. For the face-centred cubic lattice, $P(800)$ increases from about 4×10^{-3} at $\phi = 0$ to about 0.1 at $\phi = 0.125$ and to about 0.4 at $\phi = 0.15$. Similarly the value of α increases to about 1.6×10^{-4} at $\phi = 0.125$. Knots other than trefoils become more common at larger values of ϕ and the incidence of composite knots increases.

Recently, Koniaris and Muthukumar [37, 38] have carried out rather similar Monte Carlo calculations on a continuum model of a ring polymer. In their model, the polymer is represented as a set of n hard balls (representing monomers), each of radius $r \leq 1/2$, joined by unit length rods. Pairs of balls cannot overlap but otherwise the rods are freely hinged. The interactions in this model are always repulsive but decreasing the value of r corresponds to decreasing the quality of the solvent. They calculate the knot probability $P(n)$ and find their results are well represented by eqn (5.1) with α increasing as r decreases. For $r = 0.499$ they find that α is about 1.25×10^{-6}, which is rather similar to the value found for the lattice calculation with $\phi = 0$.

6 Dimensions of knotted polygons

A quite different set of questions (which at the moment can only be attacked using Monte Carlo methods) is associated with the dimensions of polygons, as a function of the knot type. If we return to the self-avoiding walk problem an obvious quantity to look at is the mean-square end-to-end length of walks with n-steps, $\langle R_n^2 \rangle$. Although there are some beautiful results on the behaviour of this quantity in high dimensions [25], little is known rigorously in three dimensions. To get some idea of the difficulty of this problem, see [39].

For many years polymer chemists and physicists have characterized polymers in dilute solution by measuring such properties as their mean-square radius of gyration. (It turns out that this can be obtained from the results of a light scattering experiment [40].) The usual assumption is that the mean-square radius of gyration of a self-avoiding walk with n-steps, $\langle S_n^2 \rangle$, behaves as

$$\langle S_n^2 \rangle = An^{2\nu}(1 + B/n^{\Delta} + \cdots) \tag{6.1}$$

where ν is believed to be about 0.59 in three dimensions, and this is expected to be a good model for the dimensions of linear polymers in good solvents.

For a polygon with n edges, in Z^3, the mean-square radius of gyration is also expected to be given by (6.1). Indeed the values of ν and Δ are thought to be the same as for a self-avoiding walk, though the parameters A and B are expected to be different. If we consider the set of n-gons which are unknotted, compared to the set of n-gons which are trefoils, we would expect that the unknots would have larger radii of gyration than the trefoils. (It is presumably this size difference which allows circular DNA molecules with different knot types to be separated by gel electrophoresis.) More generally, if we consider polygons with fixed knot type τ, how do the parameters A, ν, B and Δ in (6.1) depend on τ?

To approach this question using Monte Carlo methods we need a somewhat different kind of algorithm. First, we need to be able to sample polygons with fixed knot type. Second, it is convenient to sample polygons with different values of n in the same computation. This suggests using a "grand-canonical" approach in which n can vary. The idea is to define a Markov chain on the set of *all* polygons, so that transitions can occur between polygons with different values of n. Several such algorithms are known. One of these (known as the BFACF algorithm) employs "local" moves which do not change the knot type of the polygon. Clearly the Markov chain is not ergodic but, for this calculation, this is just what we require, since it can be shown [41] that the ergodic classes of the Markov chain are the knot classes of the polygon. That is, if the initial polygon in a realization is of type τ then (i) all other polygons in the realization will also be of type τ and (ii) every polygon of type τ has non-zero probability of occuring in the realization. (The idea of the proof is to show that Reidermeister moves can be realized by the BFACF transitions.)

Unfortunately the resulting Markov chain has slow convergence to its limit distribution (essentially because the local moves do not explore the space very efficiently). To make this into a useful method, the BFACF moves have been combined with occasional pivot moves [42]. Now the pivot moves can change the knot type so, every time a pivot move might be accepted, a check has to be made (e.g. by calculating the Alexander polynomial) that the knot type hasn't changed. If the move would change the knot type, then the resulting polygon is rejected, and the current polygon becomes the next state in the realization of the Markov chain.

This hybrid approach has been used [42] to investigate the dimensions of n-gons in Z^3 with various fixed knot type. As expected, for fixed n the mean dimensions decrease as the knot type becomes more complex. The most interesting result was that the exponent ν and the amplitude A in

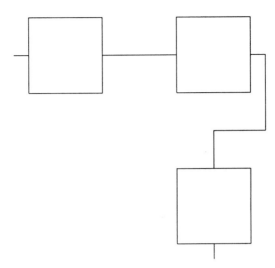

Figure 4: A pattern consisting of three knotted arcs

(6.1) seem to be independent of knot type. All the differences between knot types seem to be contained in B, the amplitude of the confluent correction.

7   Entanglement complexity of walks

A self-avoiding walk cannot be knotted. That is, for every self-avoiding walk there is a homeomorphism of R^3 to R^3 which takes the walk into the standard 1-ball. In spite of this, it is quite clear that different walks are *self-entangled* to different extents. In this section we take a first look at how one might characterise this self-entanglement.

One approach is to make use of the idea of knotted arcs, introduced in Section 3. We could define a walk as *entangled* if it contained at least one knotted arc, and measure its *entanglement complexity* as the sum of the entanglement complexities of the knotted arcs contained in the walk. This focusses entirely on local entanglements and may seriously underestimate the entanglement phenomenon.

An alternative scheme [43] is to convert the walk into a polygon and to use an entanglement complexity measure for the polygon to characterise the entanglement complexity of the walk. The obvious idea would be to add the line segment joining the first and last vertices of the walk. This doesn't work because this line segment will, in general, pass through lattice vertices and the resulting closed curve may not be simple. This problem can be avoided by offsetting the line segment by a small amount in a chosen direction, and averaging over these directions.

Since the added line segment does not in general lie in Z^3 it can pass through the associated 3-ball of a knotted arc, and a walk containing a knotted arc can be converted to an unknotted polygon by this construction. We define an *entanglement number* ζ of a self-avoiding walk as follows. For a given direction $\hat{x}$ with an infinitesimal offset we obtain a polygon which we call $\mathcal{P}(\hat{x})$ and we associate an indicator function $\chi(\hat{x})$ with this polygon which is 1 if the polygon is knotted, and zero otherwise. We define the entanglement number of the walk as

$$\zeta = \frac{\int \chi(\hat{x}) d\hat{x}}{\int d\hat{x}}. \tag{7.1}$$

What can be proved about this measure of entanglement complexity? There are two basic results which we state and prove below:

Theorem 7.1 *All except exponentially few sufficiently long self-avoiding walks have entanglement number equal to one.*

Proof: Once again the proof relies on Kesten's pattern theorem [28]. The idea is to construct a pattern consisting of three knotted arcs, such that no straight line can be drawn which passes through more than two of the associated 3-balls. This arrangement is sketched in Figure 4. Kesten's theorem asserts that this pattern occurs on all but exponentially few self-avoiding walks. For each walk which contains this pattern, no matter which line segment is added to the walk to form a polygon, the polygon will contain at least one knotted arc whose 3-ball does not intersect the line segment, and the polygon is therefore always knotted. Hence $\zeta = 1$.

The second theorem deals with the relative degree of entanglement of walks and polygons with the same value of n.

Theorem 7.2 *If c_n^o is the number of n-step self-avoiding walks with $\zeta < 1$, then*

$$\frac{c_n^o}{c_n} \geq e^{-(\kappa - \kappa_o)n + o(n)}. \tag{7.2}$$

Proof: The set of n-walks with $\zeta < 1$ includes the set of n-walks with $\zeta = 0$ and for which the two vertices of degree 1 are unit distance apart. Every (rooted, directed) unknotted polygon with $n + 1$ edges can be converted to a walk of this type by deleting the last edge. Hence

$$c_n^o \geq 2(n + 1)p_{n+1}^o = e^{\kappa_o n + o(n)}, \tag{7.3}$$

where we have made use of (3.2). Since $c_n = e^{\kappa n + o(n)}$, (7.3) implies (7.2).

8   Discussion

Polygons on a three dimensional lattice are a useful model of ring polymers. In particular, the self-avoiding condition mimics the excluded volume effect and the fact that the polygon is a simple closed curve allows us to investigate knotting phenomena. We have discussed some rigorous results which show that the probability that a randomly chosen polygon with n edges is knotted goes to unity exponentially rapidly as n goes to infinity. In addition, the same kind of argument can be used to show that almost all sufficiently long polygons contain composite knots, and that the entanglement complexity (as measured for instance by the crossing number) goes to infinity as n goes to infinity.

If we examine graphs which are more complicated than polygons then similar arguments can be used to show that embeddings of graphs with no cut edges are almost always knotted, as the number of edges in the graph goes to infinity. The proof relies on showing that cycles of the graph contain local knots. We have also examined the probability of knotting when we do not allow these local knots. There the behaviour is quite different. The numbers of unknotted and almost unknotted embeddings grow at the same exponential rate.

We have also disussed some Monte Carlo results on the incidence of knots in polygons of moderate length. Even for polygons with of the order of 1000 edges the knot probability is quite low. This implies that knotting is uncommon in relatively short polymers in good solvents. However, as the solvent quality is decreased the incidence of knots increases dramatically and it is clear that their presence must be accounted for in any model of crystallization from a poor solvent.

Monte Carlo methods can also be used to investigate the dimensions of knotted polygons, and a question of particular interest is how the radius of gyration depends on the knot type of the polygon. As expected, knotted polygons turn out to be more compact than unknotted polygons, but the critical exponent and dominant amplitude appear to be independent of knot type.

There are also interesting questions about the entanglement of linear polymers, modelled here as self-avoiding walks. We discuss one way in which the entanglement complexity of a self-avoiding walk can be defined and quantified, and prove some theorems about this measure of the entanglement complexity.

Topological effects such as knots can be important in determining the properties of polymers in various environments and we still understand rather little about these effects. Many situations have not yet been examined. For instance, how does the knot probability change if the polymer is confined to lie in a capillary? How does it depend on polymer concentration? There are many more questions than answers!

Acknowledgements

The author is grateful to Alan Sokal, Chris Soteros, De Witt Sumners and Buks van Rensburg for many helpful conversations, and to NSERC of Canada for financial support.

References

[1] H.L. Frisch and E. Wasserman, *Chemical Topology*, J. Am. Chem. Soc. **83** (1968), 3789-3795.

[2] M. Delbruck, *Mathematical Problems in the Biological Sciences* AMS, Providence, RI, 1962 pp. 55.

[3] A.V. Vologodskii, A.V. Lukashin, M.D. Frank-Kamenetskii and V.V. Anshelevich, *The knot probability in statistical mechanics of polymer chains*, Sov. Phys.-JETP **39** (1974), 1059-1063.

[4] J.P.J. Michels and F.W. Wiegel, *On the topology of a polymer ring*, Proc. Roy. Soc. A **403** (1986), 269-284.

[5] D.W. Sumners and S.G. Whittington, *Knots in self-avoiding walks*, J. Phys. A: Math. Gen. **21** (1988), 1689-1694.

[6] N. Pippenger, *Knots in random walks*, Disc. Appl. Math. **25** (1989), 273-278.

[7] C.E. Soteros, D.W. Sumners and S.G. Whittington, *Entanglement complexity of graphs in Z^3*, Math. Proc. Camb. Phil. Soc. **111** (1992), 75-91.

[8] S. Kinoshita, *On elementary ideals of polyhedra in the 3-sphere*, Pacific J. Math. **42** (1972), 89-98.

[9] S. Kinoshita, *On elementary ideals of projective planes in the 4-sphere and oriented θ-curves in the 3-sphere*, Pacific J. Math. **57** (1975), 217-221.

[10] N. Madras, A. Orlitsky and L.A. Shepp, *Monte Carlo generation of self-avoiding walks with fixed endpoints and fixed length*, J. Stat. Phys. **58** (1990), 159-183.

[11] E.J. Janse van Rensburg, S.G. Whittington and N. Madras, *The pivot algorithm and polygons*, J. Phys. A: Math. Gen. **23** (1990), 1589-1612.

[12] M. Lal, *Monte Carlo computer simulation of chain molecules*, Molec. Phys. **17** (1969), 57-64.

[13] N. Madras and A.D. Sokal, *The pivot algorithm: A highly efficient Monte Carlo method for the self-avoiding walk*, J. Stat. Phys. **56** (1988), 109-186.

[14] E.J. Janse van Rensburg and S.G. Whittington, *The knot probability in lattice polygons*, J. Phys. A: Math. Gen. **23** (1990), 3573-3590.

[15] L.H. Kauffman, *Knots and Physics* (World Scientific, 1991).

[16] D. Rolfsen, *Knots and Links* (Publish or Perish Inc., 1976).

[17] G. Burde and H. Zieschang, *Knots* (de Gruyter, 1985).

[18] O. Ore, *Theory of Graphs*, AMS Colloquium Publications Volume XXXVIII 1962.

[19] J.M. Hammersley and D.C. Handscomb, *Monte Carlo Methods* (Methuen, 1964).

[20] J.M. Hammersley, *Percolation processes II The connective constant*, Proc. Camb. Phil. Soc. **53** (1957), 642-645.

[21] D.S. McKenzie, *Polymers and Scaling*, Physics Reports **27C** (1976) 35-88.

[22] P.G. de Gennes, *Exponents for the excluded volume problem as derived by the Wilson method*, Phys. Letters **38 A** (1972) 339-340.

[23] R.G. Bowers and A. McKerrell, *An exact relation between the classical n-vector model ferromagnet and the self-avoiding walk problem*, J. Phys. C: Solid State Phys. **6** (1973) 2721-2732.

[24] G. Slade, *The scaling limit of self-avoiding walk in high dimensions*, Ann. Probab. **17** (1989) 91-107.

[25] T. Hara and G. Slade, *Self-avoiding walk in five or more dimensions*, Bull. (New Series) A.M.S. **25** (1991), 417-423.

[26] J.M. Hammersley, *Self-avoiding walks*, Physica A, to be published.

[27] J.M. Hammersley, *The number of polygons on a lattice*, Proc. Camb. Phil. Soc. **57** (1961), 516-523.

[28] H. Kesten, *On the number of self-avoiding walks*, J. Math. Phys. **4** (1963), 960-969.

[29] R.H. Fox, *A quick trip through knot theory* in Topology of 3-manifolds and related topics ed. M.K. Fort, Jnr. (Prentice Hall 1962).

[30] Y. Diao, N. Pippenger and D.W. Sumners, *On random knots*, unpublished manuscript.

[31] J.H. Conway and C. McA. Gordon, *Knots and links in spatial graphs*, J. Graph Theory **7** (1983), 445-453.

[32] S. Suzuki, *Almost unknotted θ_n-curves in the 3-sphere*, Kobe J. Math. **1** (1984), 19-22.

[33] A. Kawauchi, *Almost identical imitations of (3,1)-dimensional manifold pairs*, Osaka J. Math. **26** (1989), 743-758.

[34] S.G. Whittington and D.W. Sumners, *Almost unknotted embeddings of graphs in Z^3*, unpublished manuscript.

[35] C.E. Soteros, *Lattice models of branched polymers with specified topologies*, J. Math. Chem. to appear.

[36] J.M. Hammersley and S.G. Whittington, *Self-avoiding walks in wedges*, J. Phys. A: Math. Gen. **18** (1985), 101-111.

[37] K. Koniaris and M. Muthukumar, *Knottedness in ring polymers*, Phys. Rev. Letters **66** (1991), 2211-2214.

[38] K. Koniaris and M. Muthukumar, *Self-entanglement in ring polymers*, J. Chem. Physics **95** (1991), 2873-2878.

[39] J.M. Hammersley, *Long chain polymers and self-avoiding random walks*, Sankhyā **25** (1963), 29-38.

[40] H. Yamakawa, *Modern Theory of Polymer Solutions* (Harper and Row, 1971).

[41] E.J. Janse van Rensburg and S.G. Whittington, *The BFACF algorithm and knotted polygons*, J. Phys. A: Math. Gen. **24** (1991), 5553-5567.

[42] E.J. Janse van Rensburg and S.G. Whittington, *The dimensions of knotted polygons*, J. Phys. A: Math. Gen. **24** (1991), 3935-3948.

[43] E.J. Janse van Rensburg, D.W. Sumners, E. Wasserman and S.G. Whittington, *Entanglement complexity of self-avoiding walks*, unpublished manuscript.

DEPARTMENT OF CHEMISTRY, UNIVERSITY OF TORONTO, TORONTO, CANADA M5S1A1

E-mail address: swhittin@alchemy.chem.utoronto.ca

Proceedings of Symposia in Applied Mathematics
Volume **45**, 1992

Knots and Chemistry

JONATHAN SIMON

ABSTRACT. Topologically novel molecules such as linked rings, Mobius ladder graphs, and trefoil knots have been synthesized "from scratch". The structures are orders of magnitude smaller than the DNA knots, so the emphasis here is on how to control the synthesis (geometric rigidity of some parts of the molecules, flexibility of others) and how to prove the structures are what one hopes. While the topology content was originally just motivation ("Can a link be synthesized?"), one cannot prove (chemically) that some flexible molecule is a knot or link without [knowing and] invoking some topological properties of the alleged structure, e.g. chirality. By showing that flexible symmetries cannot always be discovered via rigid ones, topology also has influenced chemical methodology. More generally, chemistry and knot theory both encourage study of "knot theory of graphs", including classification, computable invariants, and symmetries.

1. Topologically novel molecules

Chemists can make knots! The search for geometrically and topologically novel structures has been a motivating force in organic synthesis for many years. The motivation seems to be a combination of aesthetics, searching for materials with novel properties, and a constant desire to extend the limits of synthetic capabilities. In 1989, the total synthesis of a molecular trefoil knot [dbs] was announced. The structure [FIGURE 1] consists of 124 atoms which may be viewed as a macro-cycle of 86 atoms with several smaller cycles added.

AMS Subject Classifications 57M25, 92E10

This paper is in final form and no version of it will be submitted for publication elsewhere.

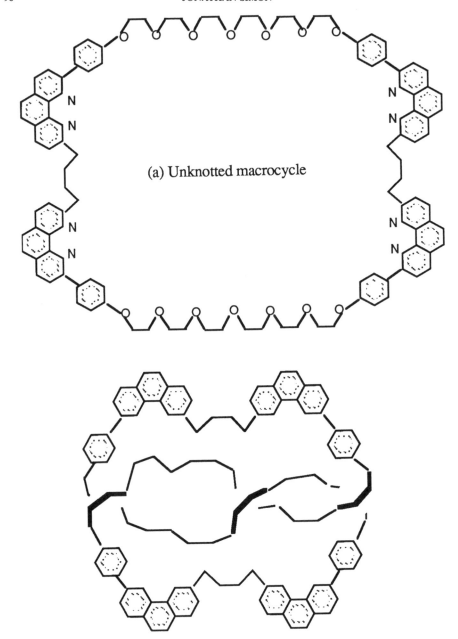

(a) Unknotted macrocycle

(b) Molecular trefoil knot

Topological Stereoisomers
FIGURE 1

The synthesis was achieved by maintaining rather strict geometric control throughout the building process. The situation here is different from DNA knots in two key ways: In the DNA situation, knots and links are obtained by having enzymes act upon relatively flexible closed macromolecule loops, allowing "strings" to pass through one another; and the structures are large enough to *see* (via electron microscope). In total organic synthesis, however, the structures are orders of magnitude smaller; the emphasis is on how exactly can the molecule be built up "from scratch" and how can one prove the result - all measures of structure are indirect (though x-ray crystallography is so trusted that one might call that direct observation of a molecule).

The knot is a culmination. Other fascinating structures achieved in the past 30 years, illustrated schematically in [FIGURE 2], include linked rings, *catenanes* [was] [sch] [sau], and Mobius bands [wal82] [wal83]. References [dbs91], [bsc], and [wal85] are enjoyable expositions of this adventure.

Where is the mathematics? What theorems about knots and links contribute to the syntheses or the proofs of structure (i.e. the chemical proofs that the stuff in the test tube really has the conjectured structure)?

The relation between knot theory and organic synthesis has been (still is) primarily that of motivation: Chemists are fascinated by elegant and unusual structures; they pursue them because molecules with novel structures will have novel properties, as a motivation for constantly challenging and expanding their synthesis techniques, and even just for aesthetic reasons. But there must be some technical connections. One cannot *prove* (chemically) that some flexible molecule has a certain topological structure without invoking some topological property of the alleged structure (or succeed in crystalizing the substance and verifying the entire geometry by x-ray crystallography). For linked rings, the "property" has just been the phenomenon of linking itself: Unbroken, one ring will pull the other along with it; once either is broken, the rings drift apart freely. For Mobius bands and knots, there also has been *chirality* - the phenomenon of an object being distinguishable from its mirror image.

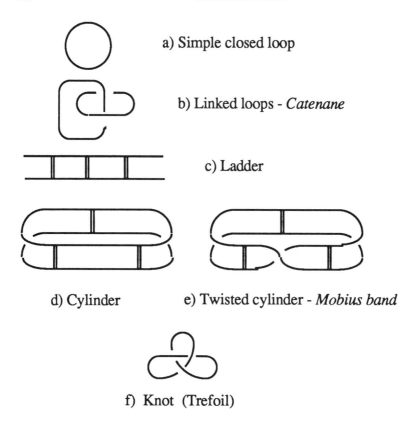

a) Simple closed loop

b) Linked loops - *Catenane*

c) Ladder

d) Cylinder   e) Twisted cylinder - *Mobius band*

f) Knot (Trefoil)

Topologically Novel Molecules
FIGURE 2

A structure is *chiral* if it cannot be moved so as to coincide with its mirror image. To make this notion well-defined, one must declare what kinds of motions are allowable. A person's left hand is chiral (that's the origin of the word) since the left and right hands are not superimposable. The usual frame of directed xy axes in the plane R^2 is chiral if we restrict to rigid motions within the plane, but if we allow rigid motions in 3-space, or if we allow free swiveling at the origin then we can twist the frame to look like its mirror image. The "right-handed" trefoil knot [FIGURE 2 (f)] is well-known to be chiral even within the topological world of complete flexibility, that is no matter how much stretching, bending, etc. is allowed, it is impossible to continuously deform the knot to its mirror image.

For chemical structures, the questions of chirality and other aspects of symmetry are generally understood in the context of considerable rigidity. Geometric properties of a structure such as bond-lengths and bond-angles (as well as the chemical identities of the various parts of the structure) are all constraints on allowable motions; however bonds do stretch (a little) and bond angles do flex (somewhat more), and even without allowing these kinds of flexibility, there still is the apparent ability to readily swivel around certain bonds. For large structures, these small flexibilities may add up to a lot, making topological analysis fruitful. If a structure is *topologically chiral* (such as the Mobius ladder molecule in [FIGURE 2 (e)]) then it must be chemically chiral, and the latter property is measurable in the lab. So when a certain molecule is measured as chemically chiral, that is a piece of the proof that it is indeed the topologically chiral trefoil knot [FIGURE 1(b)] rather than the other structure [FIGURE 1(a)] that the synthesis makes plausible. In the case of Mobius bands, more correctly *Mobius ladder graphs*, topological chirality was not known until after the chemistry was done and the question raised [wal82, wal83, sim86].

Partly in response to the chemistry, partly as a natural generalization of "classical" knot theory, there has been considerable interest recently (along with crucial earlier papers by Kinoshita, Suzuki, and Conway/Gordon) in the "knot theory of graphs" [see e.g. papers by one or groups of Flapan, Kawauchi, Kinoshita, Litherland, Scharleman, Simon,Suzuki, Thompson, Wolcott, Wu, Yamada, Zhao]. At the same time, mathematical chemists such as P. Mezey (see References [M..], [AM], [ATM])) have been exploring how chemical ideas such as chirality, symmetry, and molecular shape can be studied through topology.

Classical knot theory deals with embeddings of simple closed curves in 3-space; the questions are the standard mathematical array of invariants, symmetries,

taxonomy. One can ask exactly the same kinds of questions about graphs: Is this
spatial embedding of a K_4-graph topologically equivalent to that one [FIGURE 3]?

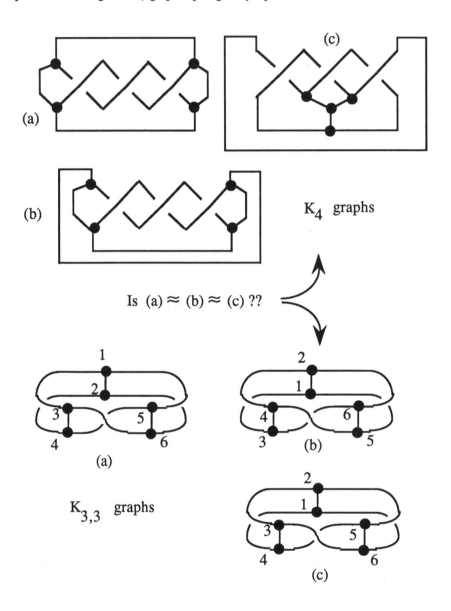

Knot Theory of Spatial Graphs

FIGURE 3

What kinds of (flexible) symmetries can a given $K_{3,3}$ -graph have [FIGURE 3]? Are there polynomial invariants that are useful for distinguishing knotted embeddings of some graph? When is a graph in space "unknotted"? How many knottings exist of theta-curve graphs with plane projections having at most k crossings?

Also, important insight has been gained [fla87] into the relation between rigid and flexible symmetry, echoing concerns in knot theory about invertible vs. strongly invertible, or amphicheiral vs. periodically amphicheiral, knots. Regardless of its own geometric symmetry or lack thereof, if a given structure A can be deformed to one B that is rigidly identical with its mirror image B^* then certainly the original A is flexibly achiral, that is there exists a pathway for deforming A to A^* . But suppose no such rigid *symmetry presentation* B exists; does this imply that A cannot be deformed to A^* ? Walba [wal83, wal85] raised this question in connection with the 3-rung Mobius ladder, [FIGURE 2(e)]. Manipulation with molecular modeling materials yielded no apparent symmetry presentation, which led to his conjecture (later verified [sim86] that the structure was topologically chiral and to the question, answered [fla87] in the negative: There exist structures (e.g. knots or knots with distinguished points) that can be deformed to their mirror images but have no topologically accessible symmetry presentations.

The analysis of rigid vs. flexible achirality is a case where knot theory has contributed not so much to a particular chemical synthesis or structure proof, but rather to chemical methodology. In another direction, it has been discovered [fla&w][lith90][sim91] that some graphs are *intrinsically asymmetric* : No matter how they are embedded in 3-space, the embeddings are chiral or do not allow some other flexible symmetries! It remains to be seen whether molecules based on intrinsically topologically asymmetric frameworks will exhibit chemical properties that echo the strong mathematical situation. This brings us back to the original interaction between topology and chemistry, that of suggesting novel targets for synthesis.

2. Routes for synthesis

Complicated structures are build up from simpler ones. The process may involve steps that are probabilistic/statistical, where a reaction produces a distribution of several products, one of which is the desired goal, or it may involve steps that are characterized by strict geometric control, where the shapes of the precursors force the

outcome structure . In fact, there usually are both elements in every step - each reaction has some probability of producing undesirable products and even statistical syntheses need to be controlled enough to give reasonable yield rates.

CATENANES In [FIGURE 4] we illustrate (schematically) several approaches to the synthesis of catenanes.

The first [f&w] [was] represents the statistical approach. Long flexible chain molecules are made to cyclize by a reaction at their ends. This is done in low concentrations to maximize ring formation and minimize polymerization. Sometimes a strand may thread through a loop before closing itself, producing linked rings. A low concentration of open strands in the presence of relatively many already closed loops would maximize the chances for linking; of course the length, thickness, and flexibility of the chains are points of control. Wasserman was able to obtain and isolate links in which the loops were (essentially) hydrocarbons with 34 carbon atoms strung end to end.

The second scheme [4(b) middle], that of Schill [sch] is characterized by rigid control. First one obtains a loop with an added crosslink connecting opposite sides. Viewing the loop as roughly planar, the next step is to attach long "arms" to the crossmember; bond angle rigidity and "steric hinderance" (part of a molecule bumping into other part) encourage one arm to run above the plane of the loop, the other below. The free ends of the arms are made to join, and then the crosslink is broken, resulting in a link of two components.

a) Simple closed loop

b) Linked loops - *Catenane*

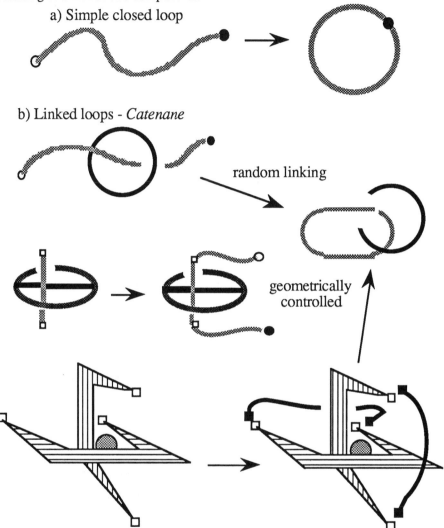

random linking

geometrically controlled

Routes for Synthesis-I. Catenanes

FIGURE 4

(Schill's catenanes - he produced many, including some of three rings - were more complicated chemically than Wasserman's but were approximately the same size.)

The third scheme [FIGURE 4(b) bottom] is the approach of Dietrich-Buchecker and Sauvage [dbs84] [sau]. Three small rigid rings are joined to form (the central part of) an "angle" (see aromatic parts of [FIGURE 1]). Two such angles form a rigid complex around a copper ion, Cu^+. The free ends of the angles are connected by chains; by controlling the lengths of the chains, dilutions, etc., the reactions produce good yields of catenane-Cu^+ complex. Finally, removing the copper gives pure catenane. As one might guess, when the chains are added, they sometimes connect one angle to another rather than to itself, so it turns out the result is 27% catenane and the rest unlocked rings or open chains joined end to end.

LADDERS AND MOBIUS BANDS We illustrate in [FIGURE 5] the scheme used by Walba [wal82] [wal83] [wal85] to produce the first molecular Mobius band. First, a simple "H" shaped structure is repeated to produce a "3-rung ladder". The 'sides' of the ladder are chains made of the same -O-C-C-O- pieces as seen in [FIGURE 1]. (This coincidence that Dietrich-Buchecker and Sauvage were using chemistry similar to what Walba had used produced some enjoyable speculation and even some laboratory effort towards joining the Sauvage "hooks" with the Walba "ladders"). The "rungs" of the ladder are carbon-carbon double bonds and so one should view the structure as pinched very closely at the rungs. Then the ends of the ladder are joined, forcing the molecule into a (possibly twisted) cylinder. As always, dilution encourages ladders to close on themselves rather than attaching to others. With 3-rung systems, two products are obtained: the untwisted cylinder (unequivocally) and (very convincingly) equal parts of left- and right- half twisted "Mobius ladders". It may be misleading to call this a Mobius <u>band</u> since the structure is not a surface, but more a (thickened) one-dimensional object.

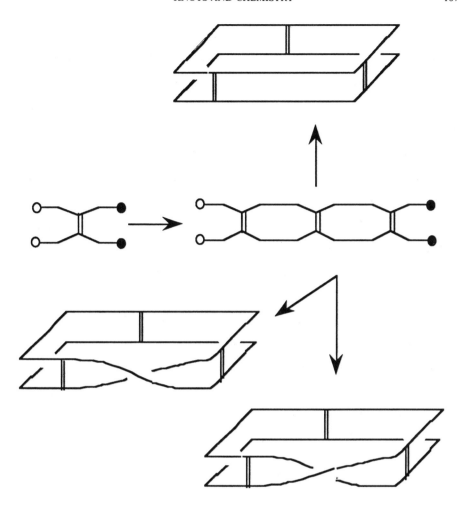

Routes for Synthesis-II. Ladders and
Mobius bands

FIGURE 5

It seems plausible that if a 3-rung ladder can produce Mobius bands then working with longer ladders can produce more twists [FIGURE 6]. This was a conjectured route to linked rings and knots [wal86]. While 4-rung systems were explored, and the cleaving of the rungs was readily available, the final yields were still heavily dominated by untwisted and one-half-twist structures; other products were evident but in very small quantities and their structure was not established. In addition to the obvious question of how many segments will give enough molecular flexibility to allow more twists, the problem apparently gets complicated by unwanted reactions between parts of the molecule that are held far apart in the smaller, 'tighter', versions but allowed to come together in longer looser systems. It is much easier for us to speculate on elegant molecular architectures than to actually make them.

THE KNOT   The total synthesis of a trefoil knot [dbs89] [FIGURE 1] was accomplished by extending the "hook" method for catenanes [FIGURE 7]. The basis is an 'S-hook' structure consisting of two rigid angles joined by a short flexible chain. Two S-hooks form a complex with two Cu+ ions as shown in [FIGURE 7 top]. Then longer chains are added to connect the ends of the hooks and finally the copper is removed. As with the catenanes, sometimes the connecting chains attach in a way to produce a knot [FIGURE 1(b)], sometimes a long unknot [FIGURE 1(a)] and sometimes other products. As one might expect, this approach produces equal amounts of left- and right- handed trefoils.

We discuss the "proofs of structure" in the next section. But it might be mentioned here that just as the synthesis of the trefoil knot relied on the chemistry developed for the hook approach to catenanes, so too the proof of structure for the knot is enhanced by the credibility established in the proofs for the catenane. If measurement XX for the catenane was substantiated by measurement YY, then we tend to believe measurement XX for the knot even if measurement YY is not possible in that setting.

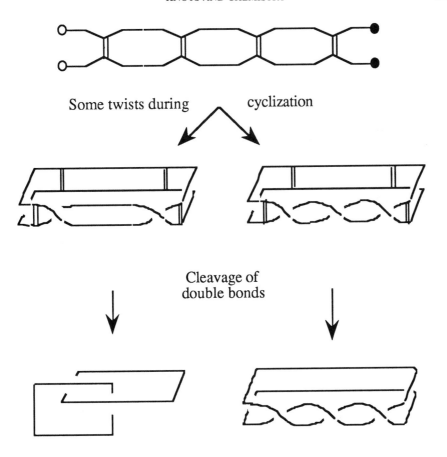

Routes for Synthesis III. Knots and Links from
Longer Ladders

FIGURE 6

Two entwined double-hooks

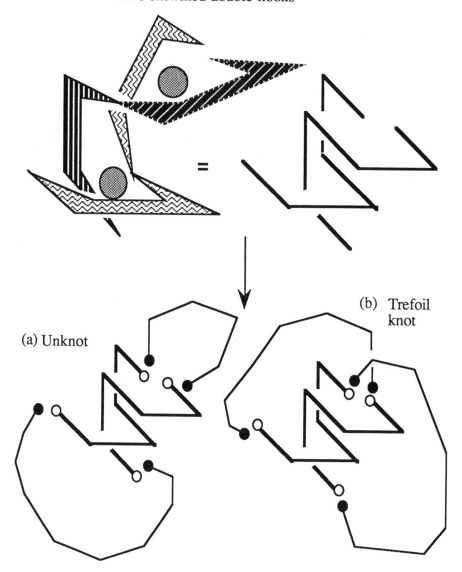

(a) Unknot

(b) Trefoil knot

=

Routes for Synthesis-IV.
Double-Hook leads to trefoil knot

FIGURE 7

3. Proofs of structure

A typical proof of structure would be that

1. *The reaction in question can only have a limited number of products.* This is based on sophisticated and elegant chemical/geometric control of the reactions and/or mechanical or computer modeling of the molecules; and

2. *Various tests indicate* (by being consistent with the hypothesis - we're dealing with laboratory science here, not mathematical proof) *that a particular product has this structure rather than that.*

REMARK. In this section, in occasional REMARKS, we also indicate some of the mathematical results, in particular on the knot theory of graphs, that have been obtained either in response to questions from chemists or just motivated by the chemical issues.

Before we go into more details for the links, ladders, and knot, we ought to discuss what is meant by *structure* as well as some of the graphical conventions commonly used to model and communicate molecular structure.

3.1. Chemical structure. Molecules have *substance, interconnections,* and *shape*. (This is a simplified, traditional description of structure, certainly arguable, but still quite useful in practice). The "stuff" is atoms; the "interconnections" are bonds between various pairs (or groups) of atoms; the "shape" is a combination of the pattern of interconnections between atoms, the geometric properties of lengths and angles that are constraints upon the bonds, and (recent addition) the topology of the embedding in 3-space. Chemical reactions of molecules and physical properties of materials depend on all levels of structure; so all should be measurable and all useful in structure proofs.

In [FIGURE 8] we illustrate a molecule made of one carbon and 4 hydrogen atoms, with bonds shown as lines. The graphical convention is that the two lines of medium density lie in the plane of the paper, the dark line C-H_b points out, and the light line C-H_a recedes. The standard geometry of a 4-valent carbon atom is that of a regular tetrahedron, with the C at the barycenter; so angle H_a-C-H_b is approximately 108°. If the four attached atoms weren't all the same (here H) then we would expect somewhat unequal angles.

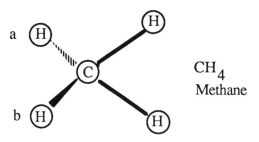

$$CH_4$$
Methane

Tetrahedral carbon
FIGURE 8

We model two molecules with the same constituent atoms in [FIGURE 9]; each has formula C_4H_{10}. But the bond patterns are different so the substances have measurably different (but similar) properties. These are called *structural* or *constitutional isomers*. Note that we are not recording the 3-dimensional geometry of the (approximately) tetrahedral bond angles and, in [FIGURE 9b] we have left the hydrogen ligands as tacitly understood.

It is quite common to leave out H atoms or other ligands in order to make the shape of the "skeleton" or "backbone" more evident. Each of the molecules in [FIGURE 10] is a ring C_6H12. The "boat" and "chair" conformations of cyclohexane remind us that the constraints of bond lengths and angles do not imply unique shapes. (In fact, trying to deduce overall shape from internal structure for large molecules, e.g. proteins or RNA, is a very active current research area. The geometric constraints

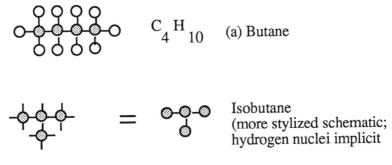

$C_4 H_{10}$ (a) Butane

= Isobutane
(more stylized schematic;
hydrogen nuclei implicit

(b) Isobutane

Constitutional isomers

FIGURE 9

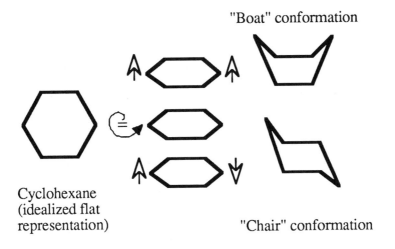

"Boat" conformation

Cyclohexane
(idealized flat
representation)

"Chair" conformation

Geometric stereoisomers

FIGURE 10

define a configuration space in which one seeks conformations that minimize energy. The tools range from plastic or metal models manipulated by hand to supercomputers.)

Structures such as (a) and (b) in [FIGURE 10] are called *stereoisomers*: They have the same atoms, same connections, but different 3-dimensional shapes. (This begins to sound like knot theory!). At room temperature, the boat and chair conformations of cyclohexane flip back and forth (which requires, hence implies, some flexibility in bond angles), so they are indistinguishable. But at sufficiently cold temperatures, they stay in position and can be told apart.

REMARK. In [ran88] [ran88'], configuration spaces for studying structures such as the cyclohexane skeleton are defined and analyzed. Randell shows that the "boat" and "chair" conformations are *topologically* distinguishable in a way that seems to account for the chemically observed flexibility of the "chair" in contrast to the relative rigidity of the "boat": *The conformations that are the "chair" have positive dimension within the configuration space, while the "boat" is 0-dimensional.*

When the chains of atoms get long, the "presenting feature" may be just the existence of a chain rather than the exact polygonal skeleton. So a molecule analogous to cyclohexane but having 30 carbon atoms in a simple chain loop might be denoted as in [FIGURE 11]. The stylized pictures of links and the knot in [FIGURE 2] represent this degree of abstraction. In the same illustration, we have preserved one chemical detail in modeling the ladders [2a,b,c]; the C-C double bonds that are the "rungs" of the ladders are distinguished from other edges of the graphs.

REMARK. The distinction between "rungs" and "sides" of twisted ladders is important. *For 3-rung systems, the Mobius ladder* [FIGURE 2e] [FIGURE 5] *is topologically chiral if we distinguish rungs from side edges* [sim86] (see *Section 3.3.2* below) *but achiral if we don't* [wal83]. *For systems with four or more rungs, the intrinsic topology of the graph, regardless of the embedding in 3-space, makes the distinction* [sim86] [wsh] and we don't need to impose it. *For four rungs and two half-twists* [FIGURE 6], *the closed ladder is topologically chiral without a priori marking of the four rungs* [sim87'].

Long chain (here closed as a loop)

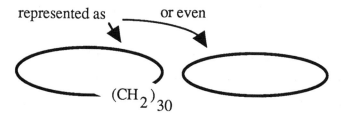

a simple loop with little or no record of the detailed structure

Graphical conventions

FIGURE 11

Here the existence of an unique pair of linking simple closed curves provides the intrinsic distinction between the rungs and sides, and the rest of the proof is simple.

If the chains are long enough then the various molecular flexibilities, in particular swiveling around C-C single bonds, may add up to enough overall flexibility that the molecule begins to behave somewhat like the objects of topology. At this point, we are abstracting molecules as "spatial graphs" [FIGURE 3]: Objects in 3-space consisting of some number of points connected by arcs. When two stereoisomers are different even at the topological extreme (e.g. [1(a)] vs. [1(b)]) of complete flexibility, we call them "*topological stereoisomers*" [wal83] [wal85].

3.2 What is measured. The arsenal of tools and cleverness of experiments used by chemists trying to verify that some organic molecule has a particular structure is astonishing. We shall describe a few of the measurements that are commonly made to verify organic syntheses.

3.2.1 *The atoms.* The mass of a molecule can be measured. From this and perhaps *a priori* information (or from other analyses), the various constituent atoms can be determined - how many atoms of each kind.

3.2.2 *Mass spectrometry.* Under relatively violent, yet not completely destructive, conditions, a molecule will cleave into (ionized) fragments in a way that is characteristic of the lower level structure - the constituent atoms and bond pattern. Mass and charge control the velocities of the fragments; so mass measurement can be refined to mass spectrometry, wherein one measures probability distributions of fragment masses. These spectra are molecular fingerprints, providing a sensitive way to distinguish structural isomers from each other. Also, they are a way to support the assertion that two molecular species that have otherwise been shown to be different are, in fact, just stereoisomers. For example, a knotted vs. unknotted loop with the same internal structure [FIGURE 1] should have identical mass spectra.

3.2.3 *NMR*. Some nuclei spin, spinning charged particles means magnetic moments, and different spins correspond to different energy levels. So if nuclei are aligned by a strong external magnetic field, and then released, they will emit energy as they revert to lower energy spins. This released energy, measured as radio waves, provides another kind of fingerprint, *nuclear magnetic resonance* spectra.

Various nuclei (^{13}C, ^{1}H are most commonly used) have characteristic signals ; but that's just the beginning. The signal shown by a certain nucleus depends on its molecular environment. Consider [FIGURE 9(c,d)], two forms of pentane. In [9(c)], there are three kinds of carbon nuclei, and NMR will reveal this clearly. The two carbons on the end (call them type "1") are each bonded to one carbon and three hydrogens; the three carbons in the middle, call them type "2", are each bonded to two carbons and two hydrogens. But there's a further distinction since one of the type 2 nuclei has only type 2 neighbors, while the other two have both 1 and 2 neighbors. So ^{13}C NMR will detect three kind of carbon nuclei - types 1 , 2 , 2' . Similarly, we expect structure [9(d)] to exhibit four types of ^{13}C nuclei, which are labeled in the diagram as 1 , 1' , 2 , 3 .

If the molecular skeleton is made of one kind of atom, as in the above example, then we are just trying to decide how many orbits there are under the automorphism group of some abstract graph.

Obviously NMR is a powerful way to distinguish structural isomers, but perhaps mass spectrometry could do the same job. However, since the signal from a particular nucleus depends not only on which others it is bonded to, but also on the 3-dimensional environment in which it lives, NMR can distinguish stereoisomers. In the two conformations of cyclohexane in [FIGURE 10], the ideal "chair" has all six carbons identical, while in the "boat" there are two different kinds. Thus if two chemical species have identical mass spectra but different NMR spectra, we have the essential ingredients for a chemical proof that they are, in fact, stereoisomers.

REMARK. We noted above that the corresponding "pure" mathematical exercise is to compute the vertex-orbits under automorphisms of a graph. Once we refine the notion of "environment" as discussed in the preceding paragraph, the corresponding mathematics then is computing vertex orbits under automorphisms induced by allowable movements in space of a graph. We are led to define [sim87] the *"topological symmetry group"* of a graph in 3-space, a topological generalization of the *Longuette-Higgins nonrigid molecular symmetry group* [loh]. Various results on topological symmetry groups of particular embeddings or on the set of all knottings of a particular graph may be found, for example, in [boy87] [fla89] [fla&w] [lith90] [sim87] [sim91]. The following are a sample.

THEOREM [boy87]. *Let* G *be a standardly embedded* K_5 *graph in 3-space; that is the 1-skeleton of a regular tetrahedron together with the barycenter and edges from the barycenter to each vertex. Then within the automorphism group of the graph (which is the symmetric group* Σ_5 *), it is precisely the even permutations that can be realized by deforming the graph in space.*

THEOREM [fla&w] [sim91]. *Let* G *be any* K_5 *graph in 3-space. Then within the automorphism group* Σ_5 *of the graph, the set of permutations that can be realized by deforming the graph in space lies within the even permutations. Similarly, there is* [lith90] [sim91] *a subgroup of index two of the automorphism group of the complete bipartite graph* $K_{3,3}$ *such that for each spatial embedding, the automorphisms realizable by motion in space lie within this subgroup.*

3.2.3 *NMR* (continued). In the structure proofs of the 3-rung cylinder and Mobius ladder [FIGURE 5], the number of different types of ^{13}C nuclei (likewise ^1H nuclei) shows that the two substances have internal structures consistent with the conjectured models. Meanwhile, NMR also reveals (through both the numbers of distinguishable types as well as the sharpness vs. fuzziness of the spectrum peaks) that one structure tends to sit in a fairly rigid conformation (the cylinder), while the other writhes more (the Mobius ladder, which mechanical models suggest should exhibit a circular corkscrew kind of motion).

REMARK. There seems to be an interesting (albeit ill posed) dynamical systems question here. The cylindrical ladder seeks an energy minimizing conformation while the nonorientable Mobius band seeks its equilibrium in the form of a periodic orbit. *Are there mathematical properties of the two systems that account for this as well as the fact that the cylinder can be crystallized while the Mobius ladder has resisted the many strategies tried so far?*

Finally, NMR helps prove chirality. Suppose a molecule can exist in two distinct mirror image forms, X and X* . Samples of pure X and X* may interact with polarized light and do so differently; this would prove the asymmetry and distinguish one from the other. But a mixture of equal parts of X and X* will behave the same as if X and X* were freely interconverting. However, if the mixture is placed in a <u>chiral</u> environment (e.g. a chiral solvent) then X and X* will respond differently, in a way detectable by NMR. From this, chirality of a species is detectable even if the reaction in which it was produced yielded equal amounts of *enantiomers* (mirror images).

3.2.4 X-RAY CRYSTALLOGRAPHY. We said earlier that measurements of structure are all indirect. One cannot see or touch one molecule. (Though for macromolecules, if one accepts electron microscopy as "seeing", then seeing is precisely how knots and links are observed. And so-called scanning/tunneling e.m. is able to "see" some of the shape of small molecules if they can be made to sit still on some substrate.) But many molecules crystallize; with the possible help of solvents or other companion atoms/molecules, they may organize into spatially regular repeated patterns. Once this happens, a characteristic response of various nuclei to incident radiation (e.g. X-ray) becomes a macroscopically measurable spatial pattern. So the precise 3-dimensional coordinates of all the nuclei in a molecule can be determined in a crystal structure. This is the *sine qua non* of structure proof. The trefoil knot was announced, with various pieces of evidence for structure, and a year later the copper-complexed form was crystallized [dgps90]. The journal in which the article on crystallization was published introduced the article saying, "There is no longer any doubt about the structure of the trefoil knotted compound that was first described in 1989. An X-ray structure analysis of the Cu-complexed compound confirms the structure derived from mass and NMR spectra."

3.3 Proofs for links, ladders, and knots. We now describe some of the highlights of the arguments for the topologically novel molecules - the catenanes, Mobius ladders, and knot. Of course, the chemical literature contains more exhaustive evidence, but we hope to indicate the key observations.

3.3.1 Catenanes. In the work of Frisch and Wasserman, [f&w] [was] the proof for catenanes was elegantly simple. There were two kinds of rings, call them S for sticky and D for detectable. The S-rings had an active site to make another reaction possible while the D-rings had a few deuterium atoms in place of hydrogens, to make them detectable by infrared spectrometry. First one kind of ring was prepared, then the other cyclization was done in the presence of the first kind of rings. By controlling dilution, the formation of hybrid links - one S-loop and one D-loop - was encouraged. The mixture was subjected to a filtering process (chromatography) in which the S-rings would stick, hence also any S-D-catenanes, and other molecules (in particular, free D-rings) would pass through. The portion that stuck (via the S-loops) was then tested for deuterium, and revealed that D-rings had been stuck as well. In this case, the topological property of the conjectured structure that was being tested was just the fact of linking; that is, the ability of one loop of a linked pair to hold or pull along the other loop despite the lack of a chemical bond between the two.

Wasserman et al did other tests, and still more were/could be done later. But at bottom, there is still the fundamental difference between *mathematical proof* and *laboratory science proof.* Tests and experiments whose outcomes are <u>consistent with</u> some hypothesis are taken as evidence for the truth of the hypothesis.

According to Schill [sch], the definitive test for catenanes is the "catenane shift" in mass spectrometry. To illustrate the principle, consider two (abstract, stylized) structures in [FIGURE 12]. Two rings of "size 5" are joined, either by a chemical bond or by linking. When the molecule is fragmented in mass spectrometry, the bonded pair should yield some fragments that have weight equal to $5 + x$, where $0 < x < 5$, i.e. a whole ring plus a proper part of the other. But in the case of linking, once one ring is broken, we expect it to drift apart from the other. In practice, the situation may be more complicated, since rings with more complicated internal structure may give up attached fragments without needing to break the ring itself. But the general principle still is that catenanes can be recognized by mass spectrometry.

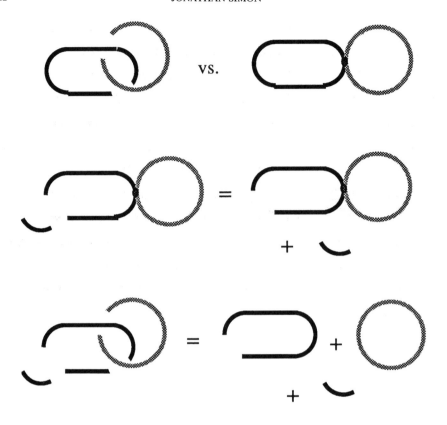

The Catenane Effect in mass spectrometry

FIGURE 12

Mass spectrometry evidence was cited in [dbs84] [sau] as persuasive for their catenanes. In fact, though, the structures were made to crystallize, both in the form of a copper-complex and in pure form without the copper center. So X-ray crystallography provided the ultimate proof, and may, in fact, be viewed as vindication of the more indirect reasoning that was also offered here, and is all that has been available in some other situations.

3.3.2 Mobius ladders. In the Mobius ladder syntheses, the key task was to distinguish the cylinder [FIGURE 5 top] from the twisted product [5 bottom] in a way consistent with the proposed structure. This was done principally via chirality. The substance believed to be the Mobius ladder was (via NMR) chemically chiral, while the other was not.

As noted earlier, manipulating hand held models indicated that the Mobius ladder could not interconvert with its mirror image (i.e. was chiral) but Walba realized that the inability of a person to find a pathway for interconversion (between mirror images, such interconversion is called *racemization*) did not imply the nonexistence of one. He raised [wal83] the conjecture that the Mobius ladder is topologically chiral, and this was subsequently established.

THEOREM [sim86]. *A standardly embedded [FIGURE 5] Mobius ladder graph in* R^3 *cannot be deformed to its mirror image in a way that sends rungs to rungs, sides to sides.*

PROOF (sketch). Form the 2-fold cyclic branched cover of the 3-sphere branched along the edge loop, J, of the ladder. Since J is unknotted, this branched covering is again the 3-sphere. The union of the three rungs in the original picture lifts to a 3-component link in the new one; and we can show that link is chiral.

It had been known for some time that a chemical structure could exist that was achiral, in the sense of allowing some pathway for racemization, but yet have no chemically feasable rigid symmetry presentation. (See e.g. Section 4 and the figure *hierarchy of achirality* in [sim87].) This would be a distressing, but apparently rare, situation for chemists (or topologists?), as the easiest way to show that some structure can be deformed to its mirror image is to deform it to a rigidly symmetric one. The phenomenon was described in [wal83] [wal85] as a "rubber glove"; some structures can be changed to their mirror images only if we subject them to great stretching in the process. Walba also observed that if we allowed infinite flexibility, then most molecules could be deformed into the plane, and a rubber glove could be deformed to a flat disk. He asked whether there exist structures that are topologically achiral but that have, even in the case of complete topological flexibility, no accessible symmetry presentation. Flapan showed [fla87] [fla88] that knot theory has such examples. The proofs depend on other results about strongly amphicheiral/invertible knots, + vs. - amphicheirality, and Smith theory.

THEOREM. *The knot* 8_{17} *can be deformed to its mirror image but it cannot be deformed to a presentation that is left setwise invariant by an orientation-reversing rigid motion of* R^3. *An appropriate satellite knot about* 8_{17} *has the same property with respect to rigid motions of the 3-sphere.*

After reading Flapan's proofs, it becomes clear that the structure shown in [FIGURE 13], a figure-8 knot with an added loop (or distinguished point), is a "topological rubber glove" in S^3.

Figure-8 knot with added loop

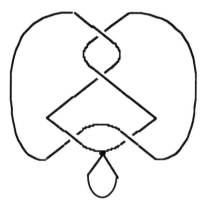

Topological "rubber glove"

FIGURE 13

Additional NMR data for the Mobius ladder is a bit more subtle. The cylinder [FIGURE 2(e) or 5 top] effectively has an inside and an outside - certain hydrogen nuclei in vs. out are distinguishable in ^{1}H NMR. But in the apparent corkscrew turning of the Mobius ladder, these protons are equivalent. So the existence of more different kinds of ^{1}H nuclei in the cylinder (which finally also was verified by X-ray crystallography, something the Mobius band has resisted) than in the "other" substance is more evidence that the "other" is the alleged Mobius band.

REMARK. This kind of analysis is part of the impetus for studying the *topological symmetry group* of a (topological model of) a molecule. This group, which was discussed in *Section 3.2.3* above, is an upper bound to feasable molecular motions, and one needs to determine the latter for a conjectured structure to help predict NMR spectra. So "pure" topology results on these groups are part of the mathematical infrastructure for future syntheses. As part of developing this infrastructure, also because it seems the natural act for a knot theorist, we have developed beginning knot tables for some simple graphs [sim87] [lith86]. So, for example, *there are exactly 10 topologically distinct embeddings of a complete graph on 4 vertices into* R^3 *having plane projections with at most 4 crossings. We have complete information regarding chirality, topological symmetries, and maximal rigid symmetry presentations for these knots.* Deleting any edge from a K_4 graph gives a graph with two vertices and three edges (properly, a multigraph) called a *theta-curve,* so we might try to analyze a knotted K_4 graph in terms of its six constituent theta-curves. *We also have tabled theta-curves up to 6 or 7 crossings (it depends on whether one considers only "prime" knottings of these graphs).* It is an accident of considering only small crossing numbers that *the ten* K_4 *graphs with at most four crossings are completely distinguished by their 6-tuples of constituent theta-curves.*

3.3.3 The trefoil knot. The knot had [dbs89] to be distinguished from the unknotted stereoisomer [FIGURE 1] that was a coproduct of the reaction that gave the knot [FIGURE 7]. Here mass spectrometry revealed that the two structures were constitutionally identical, while NMR showed they were stereochemically different. Perhaps the key observation was that the substance alleged to be the knot was (via NMR in a chiral environment) itself chiral. Also, NMR suggested a lot of molecular writhing, in contrast to relatively less for the other produce (believed to be the unknotted loop).

REMARK. Here again, the trefoil knot has a kind of "mobiosity" that seems to make it seek a dynamic, rather than static, equilibrium, much like the Mobius ladder molecule.

The final verification for the knot [dgps90] was crystallization. The knot-$(Cu^+)2$ complex was crystallized and the exact geometry verified by X-ray crystallography. The past success with catenanes and the cat-Cu+ complex lends weight to the belief that the structure of the knot would not be topologically different if the copper were removed, though apparently the uncomplexed knot has not been amenable to crystallization (which leads to yet another comparison with the Mobius ladders).

4.   Conclusion

There is none in sight. A branch of knot theory, the knot theory of graphs, is flourishing under the joint impetus of solid interactions with chemists and the joy of geometric topology. There also seems to be some good feedback into the "pure" mathematics, with people studying what might be called *physical knot theory*: What happens to classical knot theory if we endow knots with pseudo-physical properties such as thickness, stiffness, electrostatic repulsion, various other notions of energy. The papers [b&s], [f&h], [fuk], [oha], and their references, provide an entre to these questions

ACKNOWLEDGEMENTS. In order for a collaboration to develop between scientists and mathematicians, it seems necessary that one group or person learn enough of the concepts and language of the other to launch the interactions. Besides his chemical accomplishments, one must credit David Walba with being instrumental in this regard. He made initial discussions with topologists comfortable to the extent of already having well-posed topological questions and conjectures. As a bonus, his conjectures have generally (not always!) proven correct.

It also is appropriate to acknowledge the visionary encouragement provided by Ed Wegman. While he was at ONR, he was instrumental in bringing together chemists and topologists.

REFERENCES

[am91] G.A. Arteca and P.G. Mezey, "A topological analysis of macromolecular folding patterns" in *Theoretical and Computational Models for Organic Chemistry* (S.J. Formosinho et al eds.), Kluwer Academic Pub., Dordrecht, 1991, 111-124.

[am90] G.A. Arteca and P.G. Mezey, "A method for the characterization of foldings in protein ribbon models", J. Molec. Graphics 8 (1990), 66-80.

[atm] G.A. Arteca, O. Tapia, and P.G. Mezey, "Implementing knot-theoretical characterization methods to analyze the backbone structure of proteins...", J. Molec. Graphics 9 (1991), 148-156.

[boy87] J. Boyle, "Embeddings of 2-dimensional cell complexes in S^3 determined by their 1-skeletons", Topology and its Applications 25(3) (1987), 285-299.

[bsc] J. Boeckmann and G. Schill, "Knotenstrukturen in der Chemie", Tetrahedron 30, 1974, 1945-1957.

[b&s] G. Buck and J. Simon, "Knots as dynamical systems", Topology and its Applications, to appear.

[c&g] J.H. Conway and C. McA. Gordon, "Knots and links in spatial graphs", J. Graph Theory 7 (1983), 445-453.

[dbs84] C.O. Dietrich-Buchecker and J.-P. Sauvage, "Templated synthesis of interlocked macrocyclic ligands: the catenands", J. Am. Chem. Soc. 106 (1984), 3043-3045.

[dbs89] C. O. Dietrich-Buchecker and J.-P. Sauvage, "A synthetic molecular trefoil knot", Angewandte Chemie (Intl. Engl. Ed) 28 (1989), 189-192.

[dbs91] C. O. Dietrich-Buchecker and J.-P. Sauvage, "Interlocked and knotted rings in biology and chemistry", in *Bioorganic Chemistry Frontiers* Vol. 2, Springer-Verlag, Berlin - Heidelberg, 1991, 197-245.

[dgps90] C.O. Dietrich-Buchecker, J. Guilhem, C. Pascard, and J.-P. Sauvage, "Structure of a synthetic trefoil knot coordinated to two copper(I) centers", Angewandte Chemie (Intl. Engl. Ed) 29 (1990), 1154-1156.

[fla87] E. Flapan, "Rigid and nonrigid achirality", Pacific J. Math. 129 (1987), 57-66.

[fla87'] E. Flapan, "Chirality of nonstandardly embedded Mobius ladders" in *Graph Theory and Topology in Chemistry*, (R.B. King and D. Rouvray, eds.), Elsevier Sci. Pub. (Stud. Phys. Theor. Chem. No.51) Amsterdam, 1987, 76-81.

[fla88] E. Flappan, "Symmetries of knotted hypothetical molecular graphs", Discrete Applied Math. 19 (1988), 157-166.

[fla89] E. Flapan, "Symmetries of Mobius ladders", Math. Ann. 283 (1989), 271-283.

[fla&w] E. Flapan and N. Weaver, "Intrinsic chirality of complete graphs", Proc. A. M. S., to appear.

[f&h] M. Freedman and Z-X He, "On the 'energy' of knots and unknots, preprint.

[f&w] H. L. Frisch and E. Wasserman, "Chemical Topology", J. Am. Chem. Soc. 83 (1961), 3789-3795

[fuk] S. Fukuhara, "Energy of a knot" in *A Fete of Topology* (Y. Matsumoto et al eds.), Academic Press, Boston, 1988, 443-452.

[lith86] R. Litherland, "The Alexander module of a knotted theta curve", preprint 1986.

[lith90] R. Litherland, "Reeling and writhing", Seminar presentations, Univ. of Iowa, Fall 1990; paper in preparation.

[loh] H.C. Longuette-Higgins, "The symmetry groups of non-rigid molecules", Molec. Phys. 6 (1963), 445-460.

[m86] P. Mezey, "Tying knots around chiral centers...", J. Am. Chem. Soc. 108 (1986), 3976-84.

[m87] P. Mezey, "The shape of molecular charge distributions...", J. Comput. Chem. 8 (1987), 462-469.

[m90a] P. Mezey, "A global approach to molecular symmetry...", J. Am. Chem. Soc. 112 (1990), 3791-3802.

[m90b] P. Mezey, "Three-dimensional topological aspects of molecular symmetry", in *Concepts and Applications of Molecular Symilarity* (M.A. Johnson and G.M. Maggioria, eds.), Wiley, New York, 1990, 321-368.

[m91] P. Mezey, "The degree of similarity of 3D bodies: Applications to molecular shapes", J. Math. Chem. 7 (1991), 39-49.

[oha] J. O'Hara, "Energy of a knot", Topology 30 (1991), 241-247.

[ran88] R. Randell, "A molecular conformation space", *MATH/CHEM/COMP 1987, Proc. Intl. Conf. on Interfaces bet. Math., Chem., and Comp. Sci.* (R.C. Lacher ed.), *Graph Theory and Topology in Chemistry*, (R.B. King and D. Rouvray, eds.), Elsevier Sci. Pub. (Stud. Phys. Theor. Chem. No.54) Amsterdam, 1988, 125-140.

[ran88'] R. Randell, *ibid.*, 141-156.

[sau] J.-P. Sauvage, "Les catenands", Nouveau Journal de Chemie 9 (No. 5) (1985), 299-309.

[s&t89] M. Scharleman and A. Thompson, "Detecting unknotted graphs in 3-space", preprint 1989.

[sch] G. Schill, *Catenanes, Rotaxanes, and Knots*, Academic Press (Org. Chem. Mono. Ser., No. 22), 1971.

[sclf] G. Schill, G. Doerjer, E. Logemann and H. Fritz, "Untersuchungen zur Synthese von Molekulen mit Knotenstruktur...", Chem. Ber. 112 (1979), 3603-3615.

[ssfv] G. Schill, N. Schweickert, H. Fritz, and W. Vetter, "[2]-[Cyclohexatetracontan]-[Cyclooctacosan]-catenan...", Angew. Chem. 95 (1983), 909-910.

[sim86] J. Simon, "Topological chirality of certain molecules", Topology 25 (1986), 229-235.

[sim87] J. Simon, "Topological approach to the stereochemistry of nonrigid molecules" in *Graph Theory and Topology in Chemistry*, loc. cit., 43-75.

[sim87'] J. Simon, "Molecular graphs as topological objects in space", J. Computational Chemistry 8 No. 5 (1987), 718-726.

[s&w90] J. Simon and K. Wolcott, "Minimally knotted graphs in S^3", Topology and its Applications 37 (1990), 163-180.

[sim91] J. Simon, "Self-linking and the asymmetry of nonplanar graphs, in preparation.

[vls] W. Vetter, E. Logemann, and G. Schill, "Massenspektrometrische Untersuchungen einiger Catenane und Makrocyclen", Organic Mass Spectrometry 12 (1977), 351-369.

[wal82] D.M. Walba et al., "Total synthesis of the first molecular Mobius strip", J. Am. Chem. Soc. 104 (1982), 3219-3221.

[wal83] D. Walba, "Stereochemical topology" in *Chemical Applications of Topology and Graph Theory* (R.B. King, ed.), Elsevier Sci. Pub. (Stud. Phys. Theor. Chem. No. 28), 1983, 17-33.

[wal85] D. Walba, "Topological stereochemistry", Tetrahedron 41 (1985), 3161-3212.

[wal86] D.M. Walba et al., "The THYME polyethers: an approach to the synthesis of a molecular knotted ring", Tetrahedron 42 (1986), 1883-1894.

[wal87] D. Walba, "Topological stereochemistry: knot theory of molecular graphs" in *Graph Theory and Topology in Chemistry*, loc. cit., 23-42.

[wsh] D.M. Walba, J. Simon, and F. Harary, "Topicity of vertices and edges in the Mobius ladders: A topological result with chemical implications", Tetrahedron Lett. 29 No. 7 (1988), 731-734.

[was] E. Wasserman, "Chemical Topology", Scientific American 207(5) (1962), 94-102.

[wol] K. Wolcott, "The knotting of theta-curves and other graphs in S3", *Geometry and Topology: Manifolds, Varieties, and Knots* (C. McCrory and T. Shifrin, eds.), M. Dekker Publ. (Lecure Notes in Pure and Appl. Math. 105), 1987, 325-346.

DEPARTMENT OF MATHEMATICS, UNIVERSITY OF IOWA, IOWA CITY, IA 52242

E-mail: simon@math.uiowa.edu

Proceedings of Symposia in Applied Mathematics
Volume 45, 1992

KNOTS AND PHYSICS
by Louis H. Kauffman

Abstract. These notes describe the use of ideas and techniques from mathematical physics in the construction of invariants of knots and 3-manifolds.

0. INTRODUCTION

Knot theory was born from physics. In the last century, Lord Kelvin (William Thompson) theorized that atoms were vortices in the ether [201], and he comissioned Kirkman, Little and Tait to compile tables of knots in the hope of obtaining a classification of atoms. The ether soon disappeared, but the knots did not go away. They became a branch of algebraic topology, and while there were occasional glimpses of relations of knots with physics ([15],[16],[20],[29],[42],[44],[52],[75],[153], [165], [167],[174]), by and large the theory of knots and links assumed a fertile place in the domains of topology and was little influenced by the course of physical theory.

All this changed radically when Vaughan Jones discovered a new polynomial invariant of links in 1984 [79]. The Jones polynomial (see section 1) is capable of distinguising many knots from their mirror images, a feat that goes beyond the Alexander polynomial [4], known since the 1920's. Jones constructed his polynomial via a representation of the Artin braid group into a certain von Neumann algebra, an algebra that had proved structurally useful in classifying the way one von Neumann

1991 Mathematics subject classifications 57M25 (primary), 57N10 (secondary).
This paper is in final form and no version of it will be submitted for publication elsewhere.

algebra embeds in another [84].   The Jones algebra is formally identical to the
Temperley-Lieb algebra, an algebra arising in the study of the Potts model [11]. The
Potts model is a generalization of the Ising model in statistical mechanics. In this
way, comes, all at once, an interrelationship of statistical mechanics, operator algebras
and knot theory. The relationship with statistical mechanics is not Platonic. It very
soon happened [89] ,[95] ,[81],[203],[3] ,[176],[178] that one could do statistical
mechanics on a link diagram and produce new invariants of links in the process. At
the same time a number of new invariants of knots and links were discovered that
generalized   the   original   Jones   polynomial   [21],[49],
[68],[80],[88],[92],[93],[139],[173]. The first flock of these invariants are called *skein
polynomials,* because - like the Conway [30] version of the Alexander polynomial -
they are determined by simple recursive definitions on the link diagrams.

The parameters involved in doing statistical mechanics on a link diagram are for the
most part not physical, but the form of the mathematics is the same as that for the
physical theory. Certain special cases and limiting cases of physical theory apply. In
particular, a matrix relation - the Yang-Baxter equation [11] - turns out to be directly
related to the knot theory, and this relation and its solutions have been the subject of
intense study in mathematical physics.   A theory for solving the Yang-Baxter
equation evolved   under the hands of Jimbo [77], Drinfeld [37], Faddeev [39] ,
Reshetikhin [176], and other members of the Russian and Japanese schools. A device
to obtain Yang-Baxter solutions (by deforming a Lie bracket relation) became a whole
theory of deformations of Lie algebras that then generalized to Hopf algebras (Drinfeld
[37]), and has become the new field of algebra - quantum groups. It was quickly
appreciated by Jones [81], Reshetikhin and Turaev [176],[178] ,[203] and Akutsu and
Wadati [3]   that the Yang-Baxter relation and the context of quantum groups are central
to the understanding of the new invariants.

That injunction - *do statistical mechanics on a link diagram* - marks the theme of this tale. By doing mathematical physics on topological structures, all sorts of new invariants and a new way of thinking about topology has emerged. More physics than discrete statistical mechanics is involved. With Witten's work (to be described later in this introduction) came an infusion of quantum field theory, generalized Feynman integrals, conformal field theory and much more. The end is hardly in sight. These notes give a glimpse into parts of the story that are directly related to discrete statistical mechanics and the original Jones polynomial. They take the reader directly from the combinatorics of the Jones polynomial, through a version of recoupling theory for q-deformed angular momentum, to a construction of the Turaev-Viro invariant of 3-manifolds. This is one trail into a complex terrain. The Turaev-Viro invariant is one of a large number of invariants of three-manifolds that were predicted by Witten's techniques, and made combinatorial either via quantum groups [179], or by other means. The Turaev-Viro invariant is particularly interesting since it can be expressed as a partition function on the triangulation of the three-dimensional manifold.

The theme of a structure with states, and relations between these states and the topology occurs in classical knot theory (albeit in the disguise of group theory). Consider the following proof that the trefoil knot is knotted.

The First Knot

The simplest proof that the trefoil knot is knotted involves an analogue of a state of a (physical) system. Specifically, note that it is possible to color the three arcs of the trefoil diagram with three colors (red=r, blue=b, purple=p) as shown below:

The coloring is an analogue of a physical state, and the knottedness of the trefoil is seen to be reflected in the structure of state transitions as the diagram is deformed topologically.

This situation - of analyzing state change in relation to topological deformation - is closely related to the classical approach to the theory of knots (thus to the fundamental group and the Alexander polynomial). In this paper we will consider models that use the idea of summmation over all states of a given diagram. These models are analogous to partition functions for a physical system. It is a fundamental problem in the theory of knots to understand the relationship between the classical (state change under deformation) and the modern approach (partition function). The same question of can be set in the physical context. This is an important domain for the cross-channeling of ideas.

One sees that deformations (the Reidemeister moves - section 1) of the diagram allow re-coloring so that three colors are maintained and so that **each crossing has either three colors or only one color.** For example,

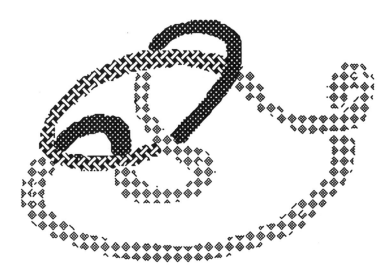

Since such a coloring can only have one color for a circle and for all unknot diagrams that are obtained from the circle by Reidemeister moves, this shows that the trefoil is knotted (i.e. it is inequivalent to the unknot via Reidemeister moves.)

The knot and link diagrams themselves are subject to a variety of interpretations. Thus we shall view the vertex

as a crossing or as a diagram for (an abstract) particle interaction - equipped with scattering matrix .

The source of this multiplicity of interpretations lies in the fact that
the diagrams themselves , viewed formally, only indicate certain patterns of relations.
Thus a crossing can be regarded as a relationship among the three arcs that are incident
to it. We can write

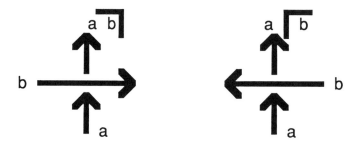

to indicate that a given arc is obtained by walking underneath another arc with left , or
right orientation. In this formalism, each arc a   acquires a dual role as **operand** , a ,
or **operator** ,

$$a \rceil.$$

One can then create an algebra (the link crystal) involving both operands and operators
that simultaneously generalizes both the fundamental group and certain forms of
coloring of the diagrams (See [88], Chapter 6 - exercises on quandles and crystals and
[107] .). There is an analogy here with the algebraic structures of quantum mechanics
where observables take on the dual roles of c-numbers (operands) and of operators.

Third Dimension

Now consider an intrinsic three dimensional viewpoint   A knot or link, seen as a
diagram, is a complex interrelation of parts that is indeed related to many other
structures such as electrical systems, graphs, and statistical mechanics. The knot or

link as an embedding in three dimensional space is a whole topological form, and it has long been the aim of toplogy to treat this form all at once, with specific decompositions being regarded as conveniences for calculation.

This magnificent conceit of topology has born extraordinary fruit in the form of many years of work with the fundamental group and the Alexander polynomial. It was John Conway in his skein theoretic approach to the Alexander polynomial who pointed out, over a hiatus of forty years (since Alexander's fundamental paper [4]) that there is a magic in the link diagrams themselves. His approach was generalized beyond belief in the wake of the discovery of the Jones polynomial [78]. And now we are presented with an extraordinary tapestry of new invariants of knots, links and three manifolds, all built combinatorially from the realm of the diagrams. Where is the invariant three-dimensional stance?

The first answer to this question came from theoretical physics. Atiyah (See[8]) suggested and Witten [215] discovered that the Jones polynomial could be described in three dimensions using the machinery of quantum field theory. This machinery depends in a fundamental way on a generalization of the Feynman path integral [41] to integrals involving integration over all fields A (gauge potentials) that can be defined on a three dimensional space. The fields take values in a Lie algebra. Each knot or link , as a specific three-dimensional embedding, becomes an observer of this field via the calculation of the trace of holonomy of the field around the knot or link. Put colloquially, what this means is that at every point x on the knot one has $B(x)= (1+A(x)dx)$, a matrix (since the field is Lie algebra valued) that can be regarded as an infinitesimal morphism from the point x on the curve K to the nearby point x+dx:

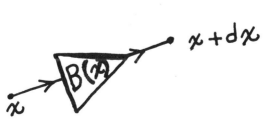

The holonomy is obtained by taking the composition of these morphisms all around the curve for some subdivision of the curve, and then taking the limit over infinitely fine subdivision. The trace of the holonomy is denoted

$$\mathrm{tr}(\mathrm{P}\textstyle\int_K \exp(A))$$

(P is for *path ordering*) , and called the **Wilson loop**. In this way, each closed loop in space contributes a complex number for a given field **A**. By taking one link (one takes the product of the Wilson loops of the components of a link) , and many fields, one can hope to average the results to obtain an invariant of the given link embedding. Witten suggested the following form of the averaging:

$$Z_K = \int DA \; \exp(L)\mathrm{tr}(\mathrm{P}\textstyle\int_K\exp(A))$$

where **L** denotes the Chern-Simons Lagrangian:

$$L = (k/4\pi)\textstyle\int_M \mathrm{tr}(A \wedge dA \; + \; (2/3)A \wedge A \wedge A).$$

The integration is taken over **M**, the three dimensional space in which the knot or link is embedded. The integral **L** is a classical integral, but the integral Z_K is a functional integral. In order to make sense out of Z_K one must first restrict the integration to certain equivalence classes of the potentials (gauge equivalence). Then one needs a measure theory on the space of potentials modulo qauge equivalence. This is not known at the present time. However, techniques of quantum field theory suggest appropriate reformulations of this integral, and in particular it is related to to the Wess-Zumino-Witten models for conformal field theories on a Riemann surface (The surface crossed with a unit integral is a model 3-space.). In this way Witten uses a series of reformulations to obtain the meaning of the invariant in terms of field theories on a surface, and these field theories turn out to embody representations of the Artin braid group in a way that matches the Jones polynomial (case of SU(2)). Thus the Jones polynomial becomes interpreted in terms of a functional integral. The fact that the mathematical analysis for functional integration is incomplete does not detract from this achievement. In fact, it leads to the question of doing what the functional integral purports to do without the analysis! That is, one wants a conbinatorial basis for the entire enterprise.

Note that in Z_K, one obtains the integral $Z(M) = \int DA \exp(L(M))$ where we emphasize the role of the 3-manifold M, and forget about the knot or link. Thus this formalism suggests the existence of invariants of 3-manifolds. Such invariants have been constructed without the functional integral by Reshetikhin and Turaev [179], Turaev and Viro [204], Kohno [125], Kuperberg [129], Kontsevich [126], Crane [33], and Walker [210]. Reshetikhin, Turaev, Viro and Kuperberg use quantum groups. Kontsevich, Khono and Crane use conformal field thoery. Walker uses a general combinatorial framework that can accept input from quantum groups or conformal field theory. In joint work [109], Sostenes Lins and the author construct the Turaev-Viro invariants using the Temperley-Lieb algebra. Lickorish [143] has used the Temperley-Lleb algebra in a different way to construct the SU(2) case of the Reshetikhin-Turaev invariant. There is an active field of work surrounding these invariants. The functional integral formula sits calmly at the center, defying our full understanding.

Quantum Three Manifolds

In the Witten picture one obtains 3-manifold invariants by forming a functional integral $Z(M) = \int DA \exp(L(M))$ where the big integration denotes the formal integration over all gauge potentials modulo gauge equivalence (for a given choice of gauge group). Suppose that the three manifold is a union of two solid handle-bodies $M = M_1 U M_2$. If M_1 and M_2 are standardly presented handlebodies, then a given surface homeomorphism $\Phi: S \longrightarrow S$ (where S is the boundary of M_1 = boundary of M_2) describes the way M_1 and M_2 are pasted together. We write

$$< M_1 \mid \Phi \mid M_2 > = Z(M),$$

and imagine that there is a Hilbert space, $H(S)$, associated to the surface S, so that whenever S is the boundary of a handlebody M_i then there is a well-defined element $<M_i|$ in this Hilbert space. Then Φ acts on the Hilbert space and $Z(M)$ is the inner product of $<M_1|$ and $\Phi|M_2>$. This idea is realized in Witten's theory through the use of the conformal field theory on Riemann surfaces and its correspondence with the Chern-Simons Lagrangian. The other approaches (mentioned above) that use conformal field theory make sense of these ideas without direct use of the functional integral. Atiyah, was the first to explicitly state axioms for invariants derived in such a pattern (generalizing Segal for conformal field theory - see [8]) , calling them **topological quantum field theories.**

Principles of Quantum Mechanics

Quantum mechanics is based on simple principles. Given a process there is associated with it an **amplitude,** Ψ, in the complex numbers. Commonly, the interpretation of the amplitude is that the product $\Psi\Psi^*$ of the amplitude and its complex conjugate is the probability of the process. The rules for computing amplitudes are as follows [41]:

1. If a process occurs in a way that can be decomposed into a set of individual steps (e.g. creations, annihilations, interactions), then the amplitude of the given process is the product of the amplitudes of the individual steps.

2. If a process may occur in several disjoint alternative ways, then the amplitude of this process is the sum of the amplitudes of the ways.

To see how these rules work, let $<a|b>$ denote the amplitude of a process that is interpreted as "starts at **a**, ends at **b**". Suppose that there is a collection of intermediate positions C (a position is not necessarily a spatial position, it connotes a specific configuration of the system) , so that the system can evolve in time from **a** to **c** and **c** to **b** for any **c** in C. Then, by the first principle, we have that the amplitude for going from **a** to **c**, **c** to **b** (a specific choice of **c**) is the product $<a|c><c|b>$. By the second principle, we have that the amplitude for going from a to b via *some* intermediate configuration c is the sum over elements of C of the products $<a|c><c|b>$:

$$\Sigma_C <a|c><c|b>.$$

Finally, if C is a *complete* list of all the intermediate possibilities, then we have, by the second principle, that

$$<a|b> = \Sigma_C <a|c><c|b>.$$

It is repeated subdivision of a path in space from point a to point b, that leads, via these principles to the Feynmann path integral. For example, if there are two intermediate states c_1 and c_2, then the amplitude becomes a double summation

$$\langle a|b\rangle = \Sigma_{C1}\Sigma_{C2}\langle a|c_1\rangle\langle c_1|c_2\rangle\langle c_2|b\rangle,$$

and if there are n intermendiate stages, it becomes an n-tuple summation

$$\langle a|b\rangle = \Sigma_{C1}\Sigma_{C2}...\Sigma_{C2}\langle a|c_1\rangle\langle c_1|c_2\rangle...\langle c_{n-1}|c_n\rangle\langle c_n|b\rangle,$$

Analytic problems occur in trying to take the limit. The heuristic arguments leading to such limits are powerful, and they give the clues for connecting classical and quantum mechanics.

Remark. Usually a quantum state is a ray in a Hilbert space, but it is convenient here to overlook this aspect. One reason for the Hilbert space is that, commonly, the amplitudes are continuous functions of space and time variables. Sums become infinite and must be handled via integration. Even in the discrete case described here, the states can be regarded as the "kets" $|b\rangle$ and their dual "bras" with amplitudes given by an appropriate inner product $\langle a|b\rangle$. Then the reversed "ket-bras" become operators whose square gives a self-multiple:

$$P = |b\rangle\langle a|$$
$$PP = |b\rangle\langle a| \; |b\rangle\langle a| \;\; = \;\; |b\rangle \; \langle a|b\rangle \; \langle a| \;\; = \;\; \langle a|b\rangle \;\; |b\rangle\langle a|$$
$$PP = \langle a|b\rangle \; P$$

The bra-ket formalism is due to Dirac [36]. Note that in this formalism we have the underlying decomposition

$$\langle a|b\rangle = \langle a| \quad |b\rangle$$

The individual bras and kets do not commute with one another, but it is assumed that $\langle a|b\rangle$ is a commuting scalar. Underneath the entire structure is the rule $|\;| = |$, a hint of Boolean algebra living in the dangerous domain of the quantum.   (If the reader will compare the Dirac formalism with the diagrammatic interpretation of theTemperley-Lieb algebra of section 1, he/she will see an interesting correspondence.)

In this Dirac context, the *completeness* of a set of intermediate states becomes the equation

$$\Sigma_X \ |x><x| \ = \ 1.$$

Then we have

$$<a|b> \qquad = <a| \ |b> = <a| \ 1 \ |b> = \ <a| \ \Sigma_X \ |x><x| \ \ |b>$$
$$= \Sigma_X <a| \ |x><x| \ \ |b>$$
$$= \ \Sigma_X \ <a|x><x|b>.$$

This is just what we deduced before from the basic principles, without the Dirac formalism.

Quantum Knots

One way to see how these principles apply to the knot theory is to consider a link with respect to a height direction in three-dimensional space. Take a plane perpendicular to this direction and let it move upward until it just intersects the link. To flatlanders living on this plane an event of creation has just begun, and momentarily a pair of particles will appear. Later, as the curves twist around each other, interactions occur. When the plane passes through a maximum in the diagram, a pair of particles is annihilated (See [135] for the view of this form of modelling in two dimensions of space, and one dimension of time (height).)

Consider an amplitude associated to a given link diagram by regarding the projection plane as 1+1 spacetime. By convention, let time run vertically up the page, and space proceed from left to right (This is the convention of the reader of English.). Position the link diagram so that it is transversal to the space levels except at critical points corresponding to maxima, minima and crossings.

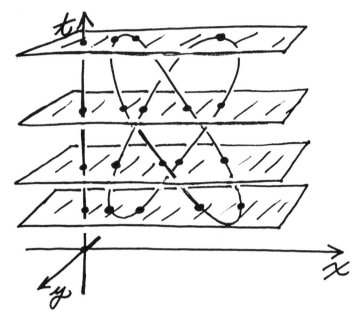

Each minimum can be regarded as a creation of two particles from the vacuum, each maximum an annihilation, and each crossing is an interaction (thought of as involving braiding in the extra spatial dimension orthogonal to the page). To each of these events we associate a matrix whose indices go over (say) the spins of the particles , and whose values are the amplitudes for each of these processes.

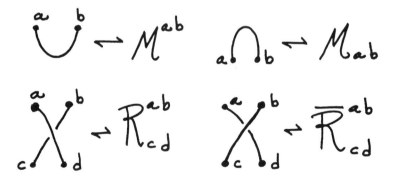

These amplitudes are a generalization of amplitudes in quantum mechanics, suitably generalized for the purposes of topology (We take leave of the usual interpretations of observation in quantum mechanics. In the topological application the amplitude itself is a real property of the system. There is no "collapse of the wave function".) The amplitude will take values in a commutative ring, and the spins will run over an arbitrary finite index set.

For a given link K, we can in this fashion calculate a vacuum-vacuum amplitude $\langle \phi|K|\phi \rangle = \langle K \rangle$ that catalogs all the possible configurations of the link as a process leading from nothing (below) to nothing (above). Of course one would like this scheme to give topological invariants. It should not depend upon the choice of direction, and it should be independent of the Reidemeister moves. See section 3 for a discussion of how this is done for the Jones polynomial and its generalizations. At the present time, essentially all known invariants of knots and links can be seen as such amplitudes, and the subject of quantum groups seems specifically designed to handle the algebra of this subject.

Remarks. It is no accident that the formula $\langle a|b \rangle = \sum_C \langle a|c \rangle \langle c|b \rangle$
is identical with the formula for matrix multiplication
$(MN)_{ab} = \sum_C M_{ac} N_{cb}$, or that matrices represent linear morphisms of a vector space or morphisms of a module over a ring. A more abstract way to put the knot theory is to say that with respect to a direction in space the link diagram becomes a composition of morphisms of modules over a ground ring. The composition starts in the ground ring (vacuum) and ends in the ground ring. A morphism (of modules over the ground ring) from the ground ring to the ground ring is just multiplication by an element of this ring , and that element is the amplitude $\langle \phi|K|\phi \rangle$.

The same remarks apply to quantum mechanics proper. One can regard an event of preparation in state a and detection in state b as a morphism from A to B, and the amplitude $\langle a|b \rangle$ as a matrix element for this morphism. This leads to speculations on generalizing everything. (Replace modules by categories, morphisms by functors.) But it is best to be circumspect at this point, lest the sirens of the categories take us against unseen rocks.

Higher   Dimensions

Of course knotting and linking goes on in higher dimensions as well. The

diagrams are quite special to the classical dimensions of knotting and linking of one-

dimensional objects in three-dimensional space. The next dimension of interest

involves surfaces embedded in four-dimensional space. We may imagine a topological

string (or membrane) theory where the membranes are embedded in four dimensional

space-time, and their three-dimensional dynamical counterparts (three-space slices of the

membrane at a particular time) are knots, links and **links with singularities**

where the singularities correspond to

1.Single points - representing the birth or death of an unknotted circle.

2.Crossing singularities - representing a saddle point of the surface in four-space.

Then all kinds of topological interactions are possible. For example, a knot and its

mirror image can interact in a standard string 4-vertex to produce two unknotted circles

(See drawing at end of this section.).

Topologically, these matters are part of the deep subject of concordance of knots and

links. There is much to be explored here topologically in analogy to the physical

theories of strings and membranes. The standard super-string theory does not , as of this

writing, participate in this topological phenomenology - due to its fascination with

higher dimensions.

 Finally, we must mention the Dirac string trick , or "belt trick" (See [88], Chapter 6

and [107] Chapters 10 & 11) - for this is one of the most well-known relationships

between knot theory and physical ideas. (A band that is twisted by 720 degrees can be

unwisted by a topological deformation that leaves the ends fixed. The same can not be

done for a 360 degree twist.)

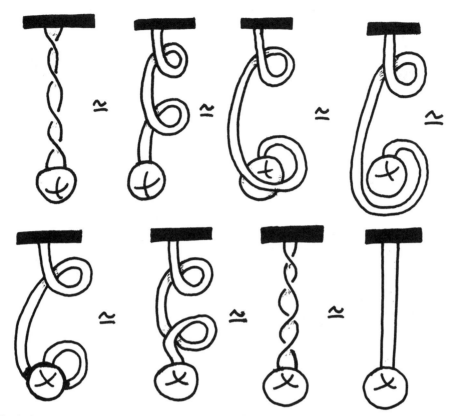

The belt trick is a direct illustration that the fundamental group of the rotation group SO(3) is of order two, and it is an analogue of the properties of spin 1/2 in quantum mechanics. In fact, as we show in [88] and [107], the string trick generalizes to a mixed mechanical/topological model of the quaternion group (actually of any subgroup of SU(2)). It is natural to ask about the relationship of the Dirac string stick and its generalizations to the link invariants and to the quantum groups. I do not know good answers to these questions. They deserve to be asked.

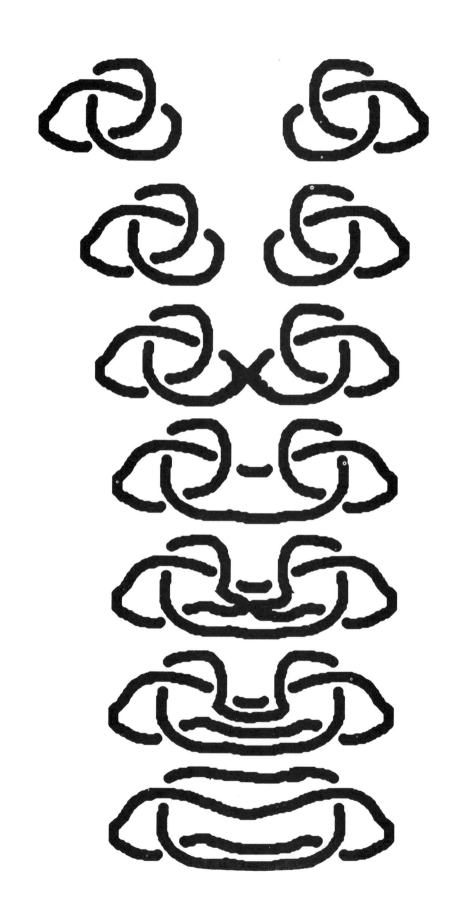

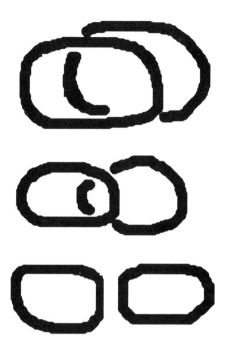

Organization

These notes are divided into six sections. The exposition spans a number of different but highly interrelated topics and techniques. In order to give the reader a picture of this investigation of knots and physics, there is a condensed exposition of the basic elements of these topics. This exposition of elements is accompanied, whenever possible by discussions and questions that motivate the author's work. This expository form is not meant to be an objective survey of the field. Rather, it is the author's continuing response to the perennial questions: What are the problems? By what method might one solve these problems?

Even so, it is not possible to do justice here to the question of methods, since a description of the sorts of interior combinatorial explorations that lie in back of this research would require composition of a book that is not yet written. This is particularly true for those parts of the project related to coloring problems and spin networks. Similarly, in the case of the use of techniques in mathematical physics, the method is borrow and try, use what can be understood. Leaps are required, and when successful these leaps are rewarding because they produce new coherent pure mathematics.

As a guide to the sections, here is a description of each section.

Section 1 discusses the bracket polynomial as a vehicle for both the Jones polynomial and the Potts model. A graphical calculus for determining the value of the three variable bracket polynomial on unlinks is formulated. This raises the question of characterization of the form of the three variable bracket for unknots and the relationship of this problem with the question whether the topological single-variable bracket detects knottedness. The latter question is equivalent to whether the Jones polynomial detects knots. The Temperley-Lieb algebra is discussed in the context of the bracket and it is shown how the Potts partition function for a plane graph can be re-expressed as a specialization of the three-variable bracket. This leads to a new result - The Potts partition function of any plane graph can be expressed as plat trace of an element of the Temperley-Lieb algebra. This in turn leads to our (joint work with H. Saleur) reformulation and generalization of the four-color problem in terms of the Temperley-Lieb algebra. In **Section 2** is reviewed the purely combinatorial theory of spin nets (using the Temperley-Lieb algebra) due to the author and S. Lins. This tangle theory provides a knot theoretic basis for the recoupling theory needed to construct the Turaev-Viro invariant. Thus we give a purely knot theoretic construction of the Turaev-Viro invariant. This work turns on delicate theorems on the behaviour of recoupling at roots of unity. One of the best questions is whether there is a generalization of these methods that will produce the oriented Reshetikhin-Turaev invariant of three manifolds as a partition function on a triangulation of the given three-manifold. Another question is to locate the combinatorial Reidemeister torsion in this description of the Turaev-Viro invariant. **Section 3** expands the context of Section 2 by introducing the concepts of Yang-Baxter models and quantum groups. It is then explained how the combinatorial spin nets of the previous section are in fact a generalization of the Penrose spin nets associated with SL(2), and that this generalization is related in a precise way to the quantum group $SL(2)_q$. With these ideas in mind, questions relating spin nets to older work of Regge, Pozano, Hasslacher and Perry show that the partition functions in the Turaev-Viro invariant are related to formalisms for Euclidean quantum gravity in 2 space and 1 time dimension. This very expanded context promises to be the right arena in which to ask questions of these invariants. One wants examples of manifolds that are not spheres, but appear as spheres to these invariants. The context of recoupling is a way to think of these questions. The old semi-classical approximations of the Regge-Ponzano work need to be generalized to the quantum group. Relations of the recoupling theory with the Chern-Simons theory are waiting to be uncovered. The section closes with a sketch a generalization of the bracket formalism to the $SL(n)_q$ models. This generalization encodes properties of the Hecke algebra and is a ground for creating a combinatorial $SL(n)_q$ recoupling theory.   **Section 4** discusses Yang-Baxter models and the (multi-variable) Alexander polynomial. By starting with the Fox free differential calculus and deriving the classical colored braid group representation, and then examining the behaviour of this braid group representation on the exterior powers of its own representation space one obtains multi-variable solutions to the Yang-Baxter equation. These Yang-Baxter solutions produce state sum models for the multi-variable Alexander-Conway polynomial that are, in turn, free-fermion models from the point of view of statistical mechanics (joint work with H. Saleur). As such they can be expressed in a *different* way as determinants. We are investigating this new determinant formulation of the Alexander polynomial. Since the classical Alexander polynomial has many beautiful properties (such as its behaviour on slice knots), the existence of rich statistical sum models for it leads to many questions of transfer. That is, one would like to transfer the proof of (say) the Fox-Milnor theorem to the statistical framework. Accomplishing this end contains strong potential for generalization, and can lead to theorems on the structure of generalized Jones polynomials on slice knots. Another avenue is to utilize the Fox-Milnor interpretation of the Alexander polynomial as Reidemeister torsion and compare with the appearance of Reidemeister torsion in the Chern Simons theory. Finally, this section raises the question of finding a generalization of the Fox calculus that could locate other Yang-Baxter solutions from the fundamental group of the knot complement. This last question is undoubtedly intertwined with the questions in **Section 5** - on the Vassiliev invariants. Vassiliev discovered how to obtain invariants of knots via the topological structure of the space of embeddings of a circle in 3-space. Birman and Lin show that the Jones polynomial and other recent invariants fit into this scheme. In this section we use the Birman-Lin result to derive explicit formulas for the Vassiliev invariants related to the Jones polynomial in terms of Tutte activities of the trees in the graph of the knot diagram (actually in terms of the Jordan Euler trails on the diagram - an equivalent formulation). This leads to a host of questions about the structure of the Vassiliev invariants and a challenge to derive these formulas directly in the original Vassiliev approach. **Section 6** discusses in a brief way the remarkable problems that come from the use of the functional integral formalism. By giving a heuristic example of this formalism the question is raised whether there is a discrete algebraic framework for these methods. I would like to understand a number of problems in this domain - particularly the nature of the skein relations, and the production of differential geometric formulas for the coefficients of link polynomials from asymptotic expansions of the functional integral. A particular challenge is the possibility of finding a framework in which (like the Fox calculus) differential calculus of the functional integral becomes a rigorous algebraic tool in relation to topological invariants.

The first appendix sketches the definitions and properties of the two-variable skein polynomials. The second appendix discusses knot epistemology, or how to make your bed in a nest of snakes.

I. JONES POLYNOMIAL AND BRACKET POLYNOMIAL

The bracket polynomial [89] is a three variable polynomial defined on link diagrams. It is useful in its own right for studying the Potts partition function and the Temperley-Lieb algebra in statistical mechanics, as a generalization of the chromatic polynomial of a plane graph, and it specializes to a state summation model for the Jones polynomial. This section will review properties of the bracket polynomial, and discuss problems in knot theory and combinatorics that are directly related to it.

We shall discuss our approach to the Turaev-Viro invariant of 3-manifolds via a combinatorial version of $SL(2)q$ quantum group recoupling theory that uses the Temperley-Lieb algebra and the topological formalism of the bracket polynomial. This gives a simple and direct approach to invariants that were originally derived either from quantum groups or from quantum field theory. In this way it is possible to raise many questions about the relationships among these fields, and at the same time remain grounded in the logic of these combinatorial and diagrammatic constructions. Many other questions about skein polynomials and Yang-Baxter models derive from looking at contexts for the bracket polynomial. Using the bracket polynomial as a guide, we venture into the surrounding terrain.

A knot or link is an embedding of a single circle or a multiplicity of circles into Euclidean three-dimensional space. Any knot or link can be projected to a plane in three-space such that the only singularities are transversal crossings of embedded curve segments, as shown below.

By standard convention, one indicates the segment that is farthest from the point of projection by drawing it with a small break. The segment nearest to the point of projection is drawn continuously. This convention for a crossing is shown below.

With this crossing convention, one can draw a **link diagram.** A link diagram is a 4-valent plane graph whose vertices have been decorated according to the crossing convention. For example, the diagram below is a diagram for the (right-handed) trefoil knot.

Any such diagram can be regarded as the projection of a link where the projection point is situated above the plane of illustration. In this way, all knots and links in three dimensional space can be indicated by diagrams.

By a theorem of Reidemeister [175], a small set of moves on diagrams suffices to articulate ambient isotopy for links embedded in three-dimensional space.

Reidemeister's moves are illustrated in Figure 1 by diagram fragments. Each fragment is assumed to be part of a larger diagram. In performing the move, the remainder of the diagram is left fixed. Along with the three moves shown in Figure 1, we also allow

deformations of the diagram that come from homeomorphisms of the plane. These deformations do not change the graphical structure of the diagram, and may be regarded as part of the background for the present discussion. (Later some of this background structure is of great importance for quantum groups.).

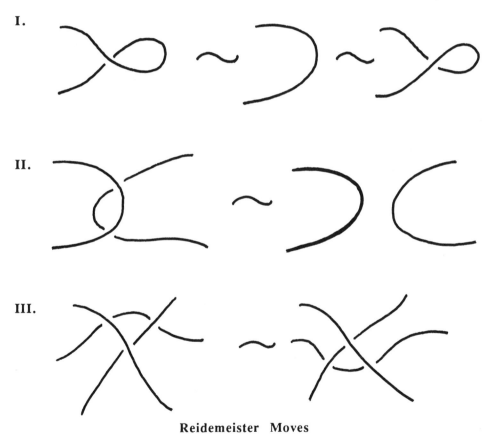

Reidemeister Moves

Figure 1

Because of Reidemeister's theorem, we say that two link diagrams are **ambient isotopic** if one can be obtained from the other by a finite sequence of Reidemeister moves and planar deformations.

Medial Construction

Before embarking on the knot theory it is useful to recall a fundamental construction that relates link diagrams with plane graphs. Given a plane graph G, there is associated with it an alternating link diagram K(G). (A link diagram is alternating if the crossings alternate under/over as one walks along the diagram.) K(G) is called the medial link diagram associated with G.

There are four regions locally incident to a crossing in an unoriented link diagram. Discriminate one pair of these regions (the A-pair) as the two regions swept out by a counterclockwise turn of the overcrossing line. Designate the remaining pair of regions by the letter B.

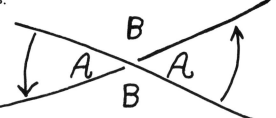

At each edge of the graph G draw a crossing so that the edge of the graph touches the A-regions of the crossing:

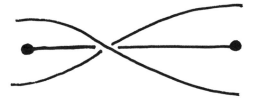

Connect the crossings together by connecting the segments adjacent to each vertex as shown below.

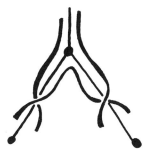

The resulting link diagram is K(G).

For example, if G is the triangle graph, then K(G) is a right-handed trefoil knot, as shown below:

The medial construction is very useful for relating constructions in graph theory with constructions in knot theory. The convention of the A and B regions is fundamental for the bracket polynomial.

Bracket Polynomial

Recall the construction of the bracket polynomial.

Given an unoriented link diagram K , one can obtain smaller diagrams by smoothing a subset of the crossings of K. Each crossing can be smoothed in two ways as shown below.

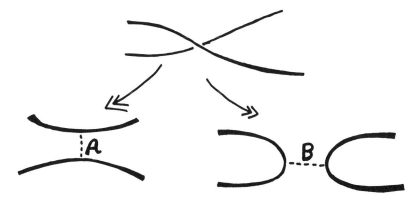

With this convention, label the two smoothings of a crossing by A or B, according as the smoothing joins the A-pair, or the B-pair. Call A and B the **vertex weights** of these smoothings. Call a **state** of the diagram K a smoothing of the

   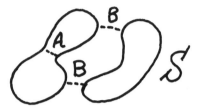

entire diagram. Note that there are 2^N states of a diagram with N crossings. A set of vertex weights is in this way associated with a given state of the diagram K. If S is a state of K, let $<K|S>$ denote the product of the vertex weights for S. It is assumed that that A and B are commuting algebraic variables so that the product of vertex weights $<K|S>$ is well-defined. Call the **norm** of a state S, denoted $\|S\|$, the number of Jordan curves in the state S. Note that any state is a disjoint collection of Jordan curves in the plane. Let d be a third algebraic variable, commuting with A and B. Define the bracket polynomial of the diagram K by the formula

$$<K> = \sum_S <K|S> d^{\|S\|}$$

where the summation is taken over all states of the diagram K.

It is then easy to see that **$<K>$** satisfies the following recursion formula

$$\left< \text{⤬} \right> = A\left< \text{)(} \right> + B\left< \text{)(} \right>$$

where the three small diagrams stand for parts of otherwise identical larger diagrams. Furthermore,   if   K is the disjoint union of a diagram K' and a disjoint curve U, then **$<K> = d<K'>$** where d is the third variable explained above. These two rules suffice to compute $<K>$ recursively on any link diagram.

The connection with topology and the Jones polynomial comes about via the following easily proved formulas:

(1) $\langle$ $\rangle = (A+Bd)\langle$ $\rangle$, $\langle$ $\rangle = (Ad+B)\langle$ $\rangle$

(2) $\langle$ $\rangle = AB\langle$ $\rangle + (ABd + A^2 + B^2)\langle$ $\rangle$

(3) $\langle$ $\rangle - \langle$ $\rangle = A(ABd + A^2 + B^2)[\langle$ $\rangle - \langle$ $\rangle]$

These formulas have a number of consequences. First of all, *they provide a graphical calculus for determining the value of the three-variable bracket polynomial on unknots and unlinks.* By definition, an unlink is a diagram that can be transformed by Reidemeister moves to a collection of disjoint Jordan curves. The formulas (1),(2),(3) give a recursive algorithm for computing bracket polynomials for unlinks. It is an unsolved problem in knot theory to characterize those diagrams that represent unknots. A corresponding problem is to *characterize the three-variable bracket polynomials of unknots (and of unlinks).* This problem about unlinks is unexplored territory in the region between knot theory and combinatorics. Information on the problem can shed light on the related problem of whether the topological bracket polynomial detects knottedness.

This brings us to the topological bracket polynomial. Call $\langle K \rangle$ the **topological bracket** when the variables are specialized to

$$B = A^{-1}$$
$$d = -A^2 - A^{-2}.$$

With this specialization, we have by (1) and (2) above that

(1') $<$⟨img⟩$> = (-A^3)<$⟨img⟩$>, \qquad <$⟨img⟩$> = (-A^{-3})<$⟨img⟩$>.$

(2') $<$⟨img⟩$> = <$⟨img⟩$>$

(3') $<$⟨img⟩$> = <$⟨img⟩$>$

Thus the topological bracket polynomial is invariant under the Reidemeister moves II and III, and it is multiplicative (i.e. it behaves via the formula (1') above) under Reidemeister move I.

One says that the topological bracket is an invariant of **regular isotopy** where regular isotopy is the equivalence relation on link diagrams generated by the second and third Reidemeister moves (and the underlying homeomorphisms of the plane). By a normalization, one obtains an invariant f_K of ambient isotopy for oriented links K:

$$f_K = (-A^3)^{-w(K)}<K>/<O>$$

where $w(K)$ is the **writhe** (or twist number) of the oriented link K and O denotes the unknot diagram.

The writhe is the sum of the signs of the crossings of K. Crossing signs are obtained by the convention shown below:

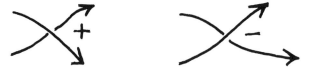

The invariant f_K is, up to a change of variable, the original Jones polynomial $V_K(t)$ [79]. In fact, $f_K(t^{-1/4}) = V_K(t)$ [89]. It is worth recalling that the Jones polynomial is completely determined by the fact that it is an invariant of ambient isotopy, that it is normalized to 1 on the unknot, and that it satisfies the skein relation shown below:

$$t^{-1}V\!\!\nearrow \quad - \quad tV\!\!\searrow \quad = \quad (\sqrt{t} - \sqrt{t^{-1}})V\!\!\longrightarrow$$

In fact, the bracket polynomial definition is very close to the original definition of the Jones polynomial in terms of von Neumann algebras. This is important for our work, and is easily seen by restricting the bracket polynomial to braids. In that context the Temperley-Lieb algebra appears in a natural way.

Braids

A **braid** is a weaving pattern formed from n distinct strands that start at a collection of n (top) points and wind around one another, always descending, arriving at a second (bottom) row of n points. See [17] for history and for formal definitions. One multiplies two braids by attaching the top of one to the bottom of the other. Taking braids up to isotopies (through braids) that fix the top and the bottom, we get the Artin Braid group. The inverse of a braid is its mirror image.

The Artin Braid Group on n strands, B_n, is generated by elementary braids

$$\sigma_1, \ \sigma_2, \ \sigma_3, \ \ldots \ \sigma_{n-1} \quad \text{and} \quad \sigma_1^{-1}, \ \sigma_2^{-1}, \ \sigma_3^{-1}, \ \ldots \ \sigma_{n-1}^{-1}$$

Each elementary braid consists in a twist between strands i and $i+1$ as shown below:

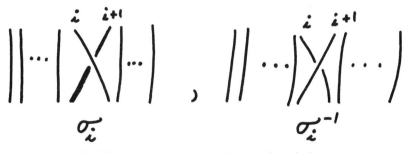

In the braid group itself, these elements are subject to the relations

$$\sigma_i \sigma_i^{-1} = 1$$
$$\sigma_i \sigma_{i+1} \sigma_i = \sigma_{i+1} \sigma_i \sigma_{i+1}$$
$$\sigma_i \sigma_j = \sigma_j \sigma_i \quad \text{if} \quad |i-j| > 1$$

These relations are a complete set of relations for the Artin Braid Group. Note that the first relation corresponds to the second Reidemeister move, the second relation corresponds to the third Reidemeister move, and the last relation is a planar deformation. In regarding braids as diagrams (subject to the three-variable bracket) we do not perform the first two relations.

Smoothing the crossing of an elementary braid σ_i yields either the identity braid (all strands parallel) or a tangle with n inputs and n outputs that we shall denote U_i as shown below. The tangle U_i consists in parallel strands from top to bottom except at the places i and i+1. There the i and i+1 inputs (top of the tangle) are connected to each other, and the i and i+1 outputs (bottom of the tangle) are connected to each other.

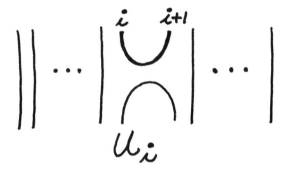

Since tangles can be multiplied just like braids (by attaching the outputs of one tangle to the inputs of the other), we can describe the **states of a braid** as tangle products of the elements U_i. For example, the product U_1U_2 for three strand braid states is illustrated below.

The value of the bracket on a braid depends upon the choice of closure of the braid. Standard braid closure is **strand closure**, obtained by attaching input i to output i by a bundle of parallel strands as shown below. Let C(b) denote the strand closure of a braid b.

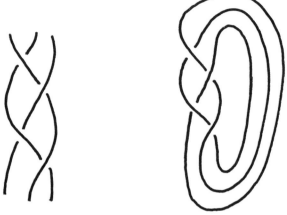

Define the bracket evaluation of the braid b to be the bracket of its strand closure: **<b> = <C(b)>**. Similarly, let T be any tangle with an equal number of inputs and outputs. Then the closure C(T) is defined just as before, and we can then define the bracket of T by the formula **<T> = <C(T)>** where C(T) is the strand closure of the tangle T. In particular, if W is a monomial in the elements U_i, then <W> is the bracket polynomial of the corresponding closed tangle. For example,

$$\left\langle U_1 U_2 \right\rangle = \left\langle \text{⋃⋂} \right\rangle = \left\langle \text{⬭} \right\rangle$$

$$= d.$$

Upon examining the multiplicative properties of the tangles, we find that it is possible for a closed component to appear in the product that is disjoint from the rest of the tangle. For example, a closed loop is seen in U_i^2, as shown below:

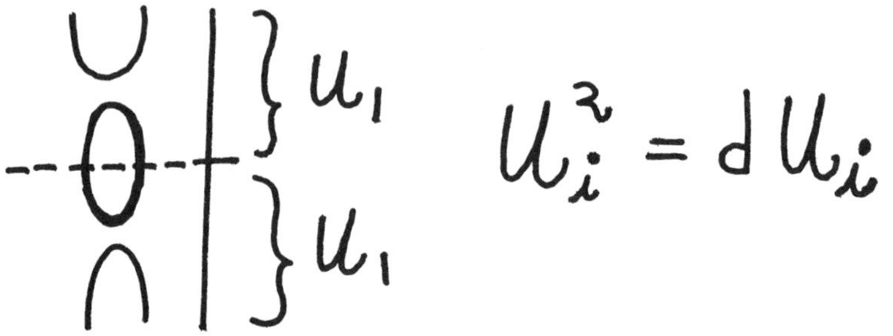

$$U_i^2 = dU_i$$

Since in bracket polynomial relations, this loop will contribute a factor of d (the third bracket variable), it makes sense to simply *identify the closed loop with d*, and to write $U_i^2 = dU_i$. We shall adopt this convention in working with the multiplicative properties of these tangles. Along with the squaring relation, above, we find

$$U_i^2 = dU_i$$

$$U_iU_{i+1}U_i = U_i, \quad U_iU_{i-1}U_i = U_i$$

$$U_iU_j = U_jU_i \quad \text{if} \quad |i-j|>1.$$

(See Figure 2.) These are the multiplicative relations in the Temperley-Lieb Algebra [11], [196]. This algebra first arose in the statistical mechanics of the Potts model, with this diagrammatic or tangle theoretic form of the algebra first appearing in [89] and [95]. The algebra itself is the free additive algebra with these multiplicative relations over the polynomial ring $Z[A,B,d]$. (Of course, for the topological bracket, we shall specialize to the ring $Z[A,A^{-1}]$ and take $d=-A^2-A^{-2}$.)

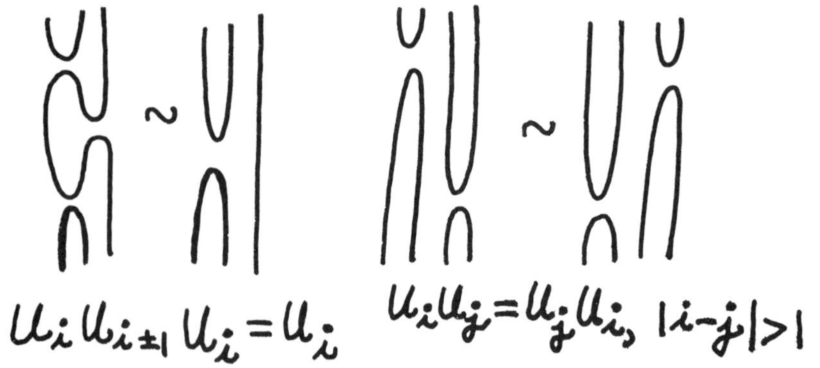

$$U_iU_{i\pm1}U_i = U_i \qquad U_iU_j = U_jU_i, \quad |i-j|>1$$

Figure 2

Let $\mathbf{T_n}$ denote the Temperley-Lieb algebra generated by
$\{1, U_1, U_2,...,U_{n-1}\}$.   It is natural to extend the category of braid-diagrams to include
elements of the Temperley-Lieb algebra, since equivalences in the Temperley-Lieb
algebra are deformations of the graphical structure of a link diagram.   Call the
diagrammatic tangles obtained by allowing products of braid generators and Temperley-
Lieb generators the **Tempered Braids, $\mathbf{TB_n}$.**   Thus a typical tempered braid is
given by the expression $\sigma_1 U_2 \sigma_1$ in TB_3.

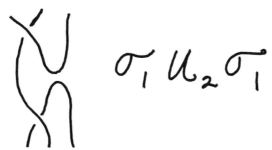

A word in σ's and U's defines a specific diagrammatic tangle, hence a specific three-
variable bracket polynomial by taking the bracket on the strand closure of that
diagram. Temperley-Lieb equivalence among the U's does not change the value of this
bracket. Braiding identities will change the value unless we specialize to the topological
bracket.

The three variable bracket is computed by a specific

Algorithm for Tempered Braids:

(1) Let w be a word in U's and σ's. Replace each occurence of an σ
or σ^{-1} in w as follows:

$$\sigma_i \text{ -----> } AU_i + B1$$

$$\sigma_i^{-1} \text{ -----> } A1 + BU_i$$

(2) Let w# be the element of the Temperley-Lieb algebra over $Z[A,B,d]$ that is obtained by multiplying out all the terms in w with these replacements. Then w# is a sum of products of the U's with coefficients in $Z[A,B,d]$.

(3) To obtain the value of $<C(w)>$ where $C(w)$ denotes the strand closure of the diagram for w, replace each product of U's in the sum in (2) by the value of the bracket on the strand closure of this element of the Temperley-Lieb algebra.

For example, let $w = \sigma_1^2$. Then $w\# = (AU_1 + B1)(AU_1 + B1)$
$= A^2U_1^2 + 2ABU_1 + B^21 = A^2dU_1 + 2ABU_1 + B^21$. Hence (Note that for n=2, $<1>$
$= d^2$), $<w> = A^2d<U_1> + 2AB<U_1> + B^2<1>$
$= A^2d^2 + 2ABd + B^2d^2$.

In the case of the topological bracket, we take $d = -A^2 - A^{-2}$ and $B = A^{-1}$. Then the correspondences in (1) of the algorithm above define a representation of the braid group to the Temperley-Lieb algebra. In this case our description of the bracket calculation can be construed as the computation of a generalized trace on a representation of the Braid Group to the Temperley-Lieb algebra. A detailed comparison with [84] shows that this description is essentially equivalent to Jones' original approach, with a combinatorial loop count replacing the algebraically defined trace function.

Plat Closure

Along with strand closure, another useful form of closure for tangles is the **plat closure**. In plat closure a tangle is given so that (as in braids and in tempered braids) the inputs are at the top and the outputs are at the bottom with respect to a height function on the diagram. Assume that the number of inputs equals the number of outputs and that this number is even. Close the diagram by adding maxima from input i to input i+1 for i odd and minima from output i to output i+1 for i odd. this closure is called the plat closure, denoted $P(T)$, the plat closure, for a given tangle T.

If T is a product of Temperley-Lieb generators U_i for U_i in T_n with n even, then T has a plat closure. Under these circumstances, define the plat trace of T, **PTr(T)**, by the formula **PTr(T)** = **<P(T)>** where <> denotes the three-variable bracket polynomial. Extend this definition linearly over Z[A,B,d] to obtain the plat trace for arbitrary elements of the Temperley-Lieb algebra.

If B is a tempered braid on an even number of strands (the number of strands of a tempered braid is the number of inputs = number of outputs) define the plat trace of B, PTr(B), by the formula PTr(B) = <P(B)>. In reference to the algorithm for tempered braids, we have that the plat trace of B is equal to the plat trace of the associated element B# in the Temperley-Lieb algebra:

PTr(B) = <P(B)> = PTr(B#).

Theorem 1.1 [103]: Every link diagram is equivalent (by planar deformations - no Reidemeister moves) to a diagram that is the plat closure of a tempered braid. Consequently, the three-variable bracket polynomial of any link diagram can be expressed as the plat trace of an element of the Temperley-Lieb algebra.

The proof of this proposition involves rearranging the maxima and minima in a link diagram, so that with respect to a given height function all internal maxima and minima are paired into local copies of U_i. The maxima and minima that then appear at the top and bottom of the diagram then form the desired plat closure of a tempered braid. The case of the trefoil knot is shown below:

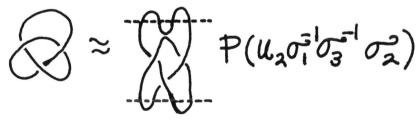

The calculation of the three-variable bracket for any link diagram via the Temperley-Lieb algebra is of independent interest for combinatorics and for statistical mechanics. For these applications, we need to explain the relationship of the bracket with the Potts model.

Potts Model and Chromatic Polynomial

The q-state Potts model in statistical mechanics has a perspicuous representation in terms of the bracket. This will be described here.

In referring to the Potts model we refer to its partition function.

A partition function in statistical mechanics is a summation, over the states of a (physical) system, of terms $\exp((1/kT)E(\sigma))$ where σ denotes a specific state, and $E(\sigma)$ is the energy of that state, T is the temperature, and k is a constant (Boltzmann's constant).

In the case of the Potts model, the physical system is represented by a graph G; the states of the system are assignments of labels from the set $\{1,2,...,q\}$ to the vertices of the graph G. Thus a state is a mapping s: Vert(G) -------> $\{1,2,...,q\}$. Call a pair of vertices of G an **edge pair** if these vertices are connected by an edge in the graph G. By definition, the energy of a state in the Potts model is the number of edge pairs such that both members of the pair receive the same color. Note that by this definition the states of energy zero in the Potts model are the **proper colorings** of the graph G with q colors. (A coloring is proper if the vertices of every edge pair receive different colors. A color is an element of the label set $\{1,2,...,q\}$.

Let $\mathbf{Z_G}$ denote the Potts partition function. Then

$$\mathbf{Z_G} = \Sigma_{\sigma}\exp((1/kT)E(\sigma))$$

where σ runs over all the states of the Potts model   (i.e. over all colorings, proper and improper, of the graph G). Note that as the temperature T goes to zero, the only contributions to the partition function are from the states where E(σ) is equal to zero. Thus *the zero temperature limit of the Potts partition function counts the number of proper colorings of the graph G.*   In fact, it is not hard to see that the Potts partition function itself satisfies recursion formulas that generalize the classical chromatic polynomial $C_G(q)$. $C_G(q)$ counts the number of proper colorings of the graph G with q colors. The result is given below ([11],[196]):

Let $v = \exp(1/kT) - 1$. Let e be an edge of G, and let G' denote the graph obtained by deleting the edge e from G. Let G" denote the graph obtained by contracting the edge e to a point. Then

$$Z_G = Z_{G'} + vZ_{G''} \qquad \text{and}$$

$$Z_{G\&p} = qZ_G$$

where G&p denotes the disjoint union of G with a single vertex p.
In the case v=-1 this recursion calculates $C_G(q)$. Z_G is a polynomial in v and q. This is often called the dichromatic polynomial.

For plane graphs G, the partition function Z_G translates to a bracket polynomial evaluation on the medial link diagram K(G) that is associated with G. The result is as follows [95]:

Let {K} denote the specialization of the three-variable bracket with

$$A = q^{-1/2}v, \ B=1, \ d = q^{1/2}. \qquad \text{Thus}$$

$$\{ \times \} = \{)(\} + q^{-1/2}v\{ \asymp \}$$

$$\{O\} = q^{1/2}.$$

Then $Z_G = q^{N/2}\{K(G)\}$ where N denotes the number of vertices in the graph G.

Since the Theorem 1.1 (stated above) implies that $\{K(G)\}$ can be expressed as a plat trace of an element of the Temperley-Lieb algebra, it follows that

Theorem 1.2 [103]. The Potts partition function Z_G can be expressed as the plat trace of an element of the Temperley-Lieb algebra for any plane graph G.

This theorem, due to the principal investigator and H. Saleur, is a generalization of the classical result expressing the Potts partition function of a rectangular lattice in terms of the Temperley-Lieb algebra. It expresses the fundamental relationship of the Potts model and the Temperley-Lieb algebra, and forms the corner stone for an on-going investigation of the Potts model and related combinatorics. In particular, we show that it follows from Theorem 1.2 and results of Jones [84] about the trace, that *there is a completely algebraic conjecture about the Temperley-Lieb algebra that is equivalent to the Four Color Theorem.* This equivalence is part of a more general conjecture about the algebra. Other conjectures about the zeroes of the chromatic polynomial for plane maps are now susceptible to analysis in terms of the Temperley-Lieb algebra.

Here is the key to our algebraic reformulation of the Four Color Theorem. In the case $q=4$, $v=-1$, we find

$$\sigma_i \longleftrightarrow R_i = 1 - (1/2) U_i$$
$$\sigma_i^{-1} \longleftrightarrow \overline{R}_i = -(1/2) + U_i .$$

Hence $R_iU_i = 0$ and $U_i\overline{R}_{i\pm1}U_i = 0$ since $U_iU_i = 2U_i$.

It is easy to see that each of the words R_iU_i and $U_iR_{i+1}U_i$ corresponds to a plat formation of   a   graphical loop.   For example:

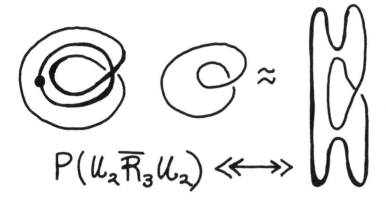

$$P(\mathcal{U}_2\overline{R}_3\mathcal{U}_2) \ll\longrightarrow\gg$$

The fact that these words are zero in the algebra corresponds to the uncolorability of a loop.   (A graph containing a loop is not vertex-colorable since the vertex at a loop is asked to be colored differently from itself.).   Call a word   W   in the   R's and U's **loop free**   if it can not be transformed   to a word containing R_iU_i or $U_i\overline{R}_{i\pm1}U_i$ by Temperley-Lieb relations among the U_i, augmented by the external relations $R_iU_j = U_jR_i$, $\overline{R}_iU_j = U_j\overline{R}_i$ for $|i-j| > 1$.

Conjecture 1.   If a word   W   (as described above)   is loop free, then W   is non-zero as an element   of the Temperley-Lieb   algebra at loop value   2.

Conjecture 2.   If a word   W   (as described above)   is loop free   and such that all indices i of the form R_i have   the   same   parity and all indices j of the form $\overline{R}_j$ have   the   opposite parity,   then   W   is non-zero as an element   of the Temperley-Lieb   algebra at loop value   2.

In [103] we show that Conjecture 2 is equivalent to the Four Color Theorem.   The parity conditions refer to what is the case for words arising from an alternating medial. To us, Conjecture 1 appears just as plausible as Conjecture 2, but Conjecture 1 is a considerable generalization of the original coloring problem.

In all these matters we have seen that the Temperley-Lieb algebra is a remarkable vehicle for translating combinatorial data about the plane into abstract algebra. There is much more to be done in this domain. In particular, by regarding the Temperley-Lieb algebra as an algebraist's substitute for the plane, there may arise new insight regarding the relationship of this algebra and the Virasoro algebra. (Compare [50],[186]. The Virasoro algebra is a central extension of the algebra of vector fields on a circle. Virasoro algebra occurs in conformal quantum field theory, and is connected in a mysterious way with the Temperley -Lieb Algebra via the continuum limit of the Potts model.)

In the next section, we detail the use of the Temperley-Lieb algebra in a link theoretic version of $SL(2)_q$ re-coupling theory.

II. Spin Networks and the Turaev- Viro Invariants

We first construct generalized antisymmetrizers .   I shall refer to the generalization as a **q-symmetrizer**. **A q-spin network** is nothing more than a link diagram with special nodes that are interpreted as these q-symmetrizers. As we shall see in section 3, the q -spin nets are generalizations of the recoupling theory of classical angular momentum.

We take $q=\sqrt{A}$. (This notation matches the q for $SL(2)q$. It is not the q of the chromatic polynomial.)

We use the bracket identity

$$\asymp = A \smile + A^{-1} \supset \subset$$

with loop value

$$\bigcirc = -A^2 - A^{-2},$$

to expand any given braid to a sum of Temperley-Lieb elements (as explained in the first section).

Now define the q-symmetrizer $\boxed{}^{|N}$ by the formula shown below. The summation is over all elements of the symmetric group S_N.

$$\boxed{}^{|N} = \frac{1}{[N]!} \sum_{\sigma \in S_N} (A^{-3})^{T(\sigma)} \boxed{\tilde{\sigma}}$$

$$[N]! = \sum_{\sigma \in S_N} (A^{-4})^{T(\sigma)} = \prod_{K=1}^{N} \left(\frac{1 - \tilde{A}^{-4K}}{1 - A^{-4}} \right)$$

Here $T(\sigma)$ is the minimal number of transpositions needed to return a permutation σ to the identity, and $\sigma^{\wedge}$ is a minimal braid representing σ with all negative crossings, i. e. with all crossings in the form shown below with respect to the braid direction

Example 1.

$$\frac{\parallel}{\parallel} = (1+\bar{A}^{4})^{-1}\left[\parallel + \bar{A}^{-3} \, X\right]$$

$$= (1+\bar{A}^{4})^{-1}\left[\parallel + \bar{A}^{3}\left[A \underset{\cap}{\cup} + \bar{A}^{-1}\right)()\right]$$

$$= \parallel - \frac{1}{(-A^{2}-A^{-2})} \underset{\cap}{\cup}.$$

Note that $f_1 = 1 - d^{-1}U_1$ is the first of a sequence of (Jones) Temperley- Lieb projectors f_n defined inductively (See e.g. [84] and [143]) via

$$f_0 = 1$$

$$f_{n+1} = f_n - m_{n+1}f_n U_{n+1}f_n$$

$$m_1 = d^{-1}, \qquad m_{n+1} = (d-m_n)^{-1}$$

$$d = -A^2 - A^{-2}$$

Theorem 2.1 [106],[109]. The Temperley-Lieb elements f_n (for loop value $-A^2$ $- A^{-2}$) are equivalent to the symmetrizers. In particular , we have the formula

$$f_{N-1} \quad = \quad \text{(diagram)}$$

Example.

$$[3]! \quad \text{(diagram)} \quad = \quad ||| \quad + \quad x \, \rangle\!\rangle| \quad + \quad x \, |\rangle\!\rangle$$

$$+ \quad x^2 \, \rangle\!\rangle\!\rangle \quad + x^2 \, \rangle\!\rangle\!\rangle$$

$$+ \quad x^3 \, \rangle\!\rangle\!\rangle \qquad (x = -A^3).$$

The Temperley-Lieb element is given by

$$\text{(diagram)} = f_2 = ||| - \left(\frac{d}{d^2 - 1}\right)\left[\begin{array}{c}\cup\\\cap\end{array}\Big| + \Big|\begin{array}{c}\cup\\\cap\end{array}\right]$$

$$+ \left(\frac{1}{d^2 - 1}\right)\left[\begin{array}{c}\cup\\\cap\end{array} + \begin{array}{c}\cup\\\cap\end{array}\right]$$

$$(d = -A^2 - A^{-2}).$$

I have emphasized the q-spin network construction for these Jones projectors. These projectors are of extraordinary use in knot theory and invariants of 3-manifolds. The Lickorish proof of existence of 3-manifold invariants [143] depends crucially on subtle properties of these operators.

q-Spin Recoupling Theory

The (Jones) trace of a tangle is calculated via the bracket polynomial by closing the tangle

One then has that $\mathrm{tr}(\mathbf{f_{n-1}}) = \Delta_n$ where Δ_n is the Chebyshev polynomial

$$\Delta_n = (t^{n+1} - t^{-n-1})/(t - t^{-1}), \qquad t = -A^2$$

Hence

The **3-vertex** is defined with q-symmetrizers:

Here a,b,c are positive integers satisfying the condition that the equations $i+j=a$, $i+k=b$, $j+k=c$ can be solved in non-negative integers.

For example

.

Note that the three-vertex is a sum of tangles, and that its insertion in a diagram means that the extended topological bracket evaluation for that diagram includes the summation over all of these diagrams in the sum of tangles.

There are various motivations for this definition of a three vertex. Some motivations come from the theory of angular momentum recoupling [167] , some come from properties of the SL(2) quantum group [99], [106], and finally the vertex is motivated by the problem of designing a vertex that can be used to create invariants of topological equivalence of three-valent graphs embedded in three dimensional space [90]. This latter motivation explains the use of braiding and symmetrizers, since one needs to construct a machinery at each vertex that will absorb braiding of the lines entering the vertex. This absorbtion is reflected in the first property of these three-vertices - multiplicativity under braiding.

By using this strand-description of the vertices, it is easy to verify that

$$(x' = x(x+2)) \qquad (\epsilon = (-1)^{\frac{a+b-c}{2}}) \qquad \Big|^c \atop a\diagup \diagdown b \ = \epsilon A^{\frac{1}{2}(a'+b'-c')} \ \Big|^c \atop a\diagup \diagdown b$$

and

$$b\bigcirc c \atop \Big|^a \ \Big|_{a'} \ = \left[\dfrac{\Theta(a,b,c)}{\bigcirc a} \right] \ \begin{array}{c} a \\ \square \\ \; \end{array} \ \delta_{a}^{a'}$$

We call the evaluation of the Θ - shaped   network

the Θ-**symbol,** and denote its value by $\Theta(a,b,c)$.

The **q-6j** symbols are defined in terms of the three-vertices by the recoupling formula.

$$\begin{array}{c} b \diagdown \quad \diagup c \\ \bullet \!-\! j \!-\! \bullet \\ a \diagup \quad \diagdown d \end{array} \ = \ \sum_i \left\{ \begin{array}{ccc} a & b & i \\ c & d & j \end{array} \right\} \begin{array}{c} b \diagdown \quad \diagup c \\ \bullet \\ | \\ \bullet \\ a \diagup \quad \diagdown d \end{array}$$

To obtain a formula for these recoupling coefficients, proceed diagrammatically:

$$\Rightarrow \begin{Bmatrix} a & b & i \\ c & d & j \end{Bmatrix} = \frac{\left[\text{(tetrahedron net)} \right]}{\text{(denominator nets)}} .$$

Here we call the factor

the **tetrahedron**. In calculating the tetrahedron, we expand its vertices to get a network of q-symmetrizers.

The final formula determines the q-6j symbol in terms of spin network evaluations of some small nets. These can be handled by combinatorial means. In the case of q=1, the various factorials in the resulting formulas are interpreted in terms of counting loop evaluations in the state expansions for the nets. In the case of general q, it is a subtle matter to obtain the analog formulas for these recoupling coefficients . Nevertheless, the exterior formalism in the q-deformed theory is identical to that of the classical theory.

This same approach can be used to prove various properties of these objects such as orthogonality relations, and pentagon (or Elliot-Biedenharn) identities that arise from the recoupling shown below:

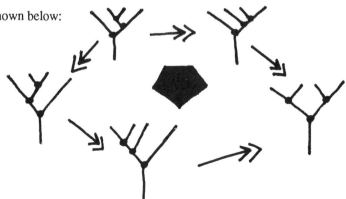

Here is a statement of these identities. Note that the Elliot Biedenharn identity involves two recoupling symbols and a sum over a product of three symbols. This corresponds to the *two and three* pattern in the pentagon recoupling that is illustrated above.

Elliot-Biedenharn Identity

$$\sum_{m} \begin{Bmatrix} a & i & m \\ d & e & j \end{Bmatrix} \begin{Bmatrix} b & c & \ell \\ d & m & i \end{Bmatrix} \begin{Bmatrix} b & \ell & \kappa \\ e & a & m \end{Bmatrix} = \begin{Bmatrix} b & c & \kappa \\ j & a & i \end{Bmatrix} \begin{Bmatrix} \kappa & c & \ell \\ d & e & j \end{Bmatrix}.$$

Orthogonality Identity

$$\sum_{i} \begin{Bmatrix} a & b & i \\ c & d & j \end{Bmatrix} \begin{Bmatrix} d & a & \kappa \\ b & c & i \end{Bmatrix} = \delta_{j\kappa}.$$

All of these evaluations work as advertised for generic q. When q is a root of unity, then the story is a bit different. We say that a triple is **r-admissible** if a+b+c<= 2r-4 where r is an integer greater than 2.

Proposition A [109]. Let q be a primitive 2r-root of unity, and

$\Theta(a,b,c)$ be the Θ-symbol for an admissible triple (a,b,c). Then (a,b,c) is r-

admissible iff $\Theta(a,b,c)$ is non-zero.

Proposition B [109]. For q a primitve 2r-root of unity the recoupling formula

for q-6j symbols exists in the sense that a vertex triple in the formula is present iff it

is r-admissible.

Under these conditions of r-admissibility, the orthogonality and Elliot-Biedenharn

identities continue to hold, and the specific formula for the q-6j symbol works since

the denominators are non-zero.

These results are the most important parts of our reformulation of recoupling theory.

They put on a strictly combinatorial footing properties that were heretofore known only

through an algebraic analysis of the quantum group $SL(2)_q$ at roots of unity. By

presenting a fully combinatorial recoupling theory, we open the way for other

constructions to be adapted or invented for this realm. In particular, this yields an

elementary approach to the invariants of Turaev and Viro for three dimensional

manifolds.

The Turaev-Viro Invariant of 3-Manifolds

We use the Matveev representation (see [151]) of three-manifolds in terms of special

spines. In such a spine , a typical vertex appears as shown below with four adjacent

one-cells , and six adjacent two-cells. Each one-cell abuts to three two cells.

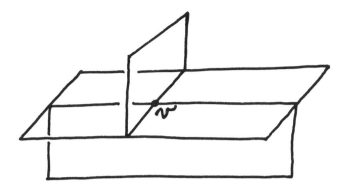

For an integer **r** >=3 the **color set** is C(r) = {0,1,2,..., r-2}.

A **state at level r** of the three-manifold M is an assignment of colors from C(r)
to each of the two-dimensional faces of the spine of M. Let q be a primitive 2r-root
of unity.

Given a state S of M, assign to each vertex the tetrahedral symbol whose edge colors
are the face colors at that vertex. The form of this assignment is shown below with the
standard orientation at the vertex .

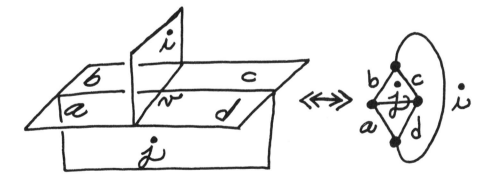

Assign to each edge the Θ-symbol associated with its triple of colors, and assign to each face the Chebyshev polynomial whose index is the color of that face.

Let $I(M,r)$ denote the sum over all states for level r, of the products of the vertex evaluations and the face evaluations divided by the edge evaluations **for those evaluations that are r-admissible (Propositions A and B).**

$$I(M,r) = \sum \frac{\prod_{v} TET(v,S) \prod_{f} \Delta_{S(f)}^{\chi(f)}}{\prod_{e} \Theta(S_a, S_b, S_c)^{\chi(e)}}$$

Here $TET(v,S)$ denotes the tetrahedral evaluation associated with a vertex v , for the state S. $S(f)$ is the color assigned to a face f, and S_a, S_b, S_c are the triplet of colors associated with an edge in the spine. $X(f)$ is the Euler characteristic of f, and $X(e)$ is equal to 1 if the edge e has graphical nodes incident to it, and is 0 otherwise.

It follows via the orthogonality and Elliot-Biedenharn identities for the q-6j symbols, that $I(M,r)$ is invariant under the Matveev moves pictured below. Hence, by Matveev's work [151], $I(M,r)$ is a topological invariant of the 3-manifold M. This is our version of the Turaev-Viro invariant.

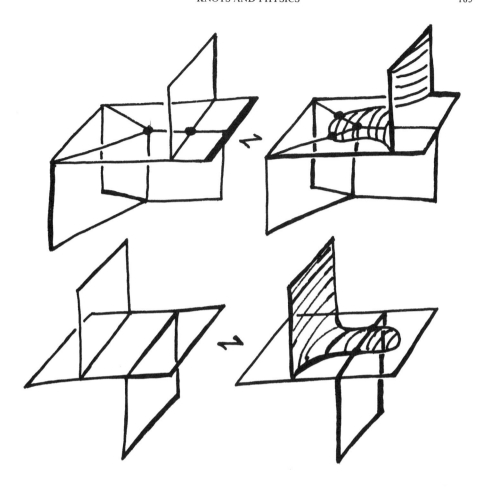

Matveev Moves

Remarks. It has been announced by Turaev [207] and by Walker [210] that the Turaev-Viro invariant is equal to the product of the Reshetikhin - Turaev invariant and its complex conjugate. Let **RT(M, r)** denote the Reshetikhin-Turaev invariant. Thus this result tells us that

$$\mathbf{I(M,r)} = |\mathbf{RT(M, r)}|^2$$

where $|x|^2$ denotes the absolute square of x. An important problem for this investigation is to *find a formula for the Reshetikhin-Turaev invariant as a finite partition function on the triangulation or special spine of the 3-manifold..* Part of our method for investigating this problem is to use the very spare and elementary treatment via the Temperley-Lieb algebra that has been outlined so far in these notes.

This completes a description of the construction and invariance of the Turaev-Viro partition function that uses only basics of combinatorial knot theory, aspects of tangles and permutations, and elements of three dimensional topology.

This approach should be compared with the methods of [143] for constructing the Reshetikhin-Turaev invariant. Lickorish defines this invariant (as do Reshetikhin and Turaev) by representing the manifold by surgery on a framed link and letting $RT(M,r)$ be a (normalized) summation of bracket polynomials (at $2r$ roots of unity) of cables $c*K$ of the link K where $c*K$ denotes a link obtained by replacing some of the components by k-fold parallel copies of that component where k ranges from 0 to r-2. In the Lickorish approach, the Jones projectors in the Temperley-Lieb algebra figure prominently as the key technical device for proving the existence and properties of the invariant. A comparison of the way in which these projectors figure in the construction of the two invariants should give insight into their common structure.

The project is open to other ideas, from quantum groups, conformal field theory, functional integrals and quantum field theory and functorial formulations of topological quantum field theories. In fact there are certain questions that demand comparison with the quantum field theory approach of Witten. His work [215] shows that the Reidemeister torsion of the three manifold is contained in averaged form in the limit of $RT(M,r)$ as r goes to infinity. The same remarks apply to $I(M,r)$. Calculations with Lens spaces [48] have shown that Witten's results are true in the surgery and conformal field theory models of these invariants. This raises an excellent problem for combinatorial topology: *Find the Reidemeister torsion directly in the surgery versions of RT(M,r) and in the partition function versions I(M,r) (for an arbitrary compact three manifold M).*

III. Yang Baxter Models and the SL(2) Quantum Group

The purpose of this section is to describe the bridge between the purely combinatorial approach to link invariants and invariants of 3-manifolds,   and methods involving quantum groups and ideas in mathematical physics. By describing this more algebraic language it becomes possible to formulate a wider class of problems.

Recall first the method (of Turaev and Reshetikhin [178]) for making invariants of links via solutions to the Yang-Baxter equation:

Associate to each diagram a height function so that the diagram is decomposed into crossings, minima (cups) and maxima (caps). Regard each of these fragments as a mapping in a category of tensor products of modules over a fixed ground ring C as described below.   In the case of the bracket, C is taken to be $Z[A,A^{-1}]$. For simplicity, let us assume that there is a single module V   associated with each component in the link diagram. Then a cup is a morphism from C to $V \otimes V$, a cap is a morphism from

$V \otimes V$ to C, and a crossing is a morphism from $V \otimes V$ to $V \otimes V$.

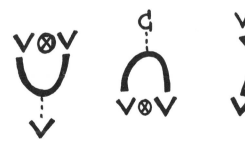
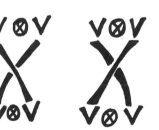

A closed link diagram arranged with respect to a height function then defines a composition of morphisms starting in C and ending in C.

Since a morphism of C as a C-module is multiplication by an element of C, we see that each link diagram with height function gives rise to a well defined element of C, the ground ring.

Given a link diagram K with a height function and a choice of morphisms for the fragments, let <K> denote the corresponding element of C. This element is to be the generalization of the link polynomial invariant, and for appropriate morphisms we shall produce the original topological bracket polynomial.

In order for <K> to be an invariant of regular isotopy, the morphisms must satisfy relations that correspond to the (generalized) Reidemeister moves for regular isotopy with respect to a given direction. These moves are shown in Figure 3.

If we represent each of the morphisms as a matrix with respect to a finite dimensional basis for the module V over the ring C, then the equations corresponding to the generalized Reidemeister moves become matrix equations. These can be written in indices with the Einstein summation convention that a repeated appearance of a given index in upper and lower position connotes the summation over that index. These matrix equations are as follows.

I. $M_{ai} \, M^{ib} = \delta_a^b \qquad M^{ai} M_{ib} = \delta_b^a$ (Cancellation)

$(\delta_b^i X_i^a = X_b^a \, , \delta_i^a X_b^i = X_b^a)$

II. $R_{ij}^{ab} \, \overline{R}_{cd}^{ij} = \delta_c^a \delta_d^b$ (Inverses)

III. $R^{ab}_{ij}R^{jc}_{kf}R^{ik}_{de} = R^{bc}_{ki}R^{ak}_{dj}R^{ji}_{ef}$ (Yang-Baxter Equation)

IV. $M_{ab}R^{bc}_{de} = \overline{R}^{cb}_{ad}M_{be}$ (Slide Move)

V. $\overline{R}^{ab}_{cd} = M_{ci}R^{ia}_{dj}M^{jb}$ (Twist Move)

It is easy to see that these equations express exactly the conditions that the morphisms corresponding to the before and after shots of the moves are identical. Hence if these matrix conditions are satisfied, then the element

$<K>$ is an invariant of regular isotopy. Since, the type three Reidemeister move corresponds to the well-known Yang-Baxter equation, it is convenient to call this model for a link invariant a **Yang-Baxter model.** To bring this procedure down to earth, here is a description of the construction of a Yang-Baxter model for the original bracket polynomial.

(I)

(II)

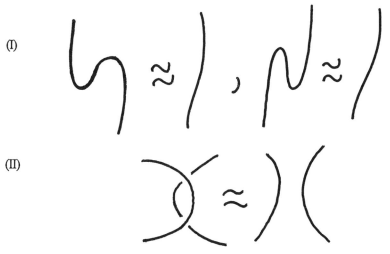

Regular Isotopy With Respect to a Direction

Figure 3   (Part 1)

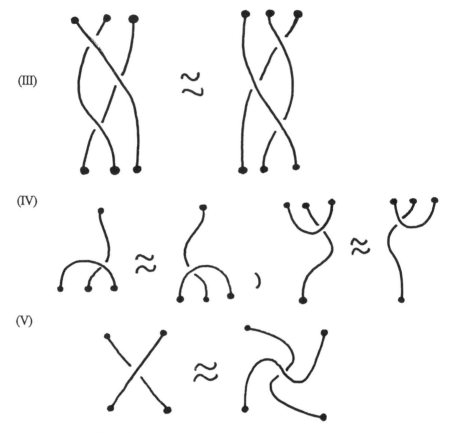

(III)

(IV)

(V)

Regular Isotopy With Respect to a Direction

Figure 3 (Part 2)

Bracket Polynomial and Yang-Baxter Equation

First recall the topological specialization of the bracket polynomial.

This bracket is defined by the equations:

1. $\langle \asymp \rangle = A\langle \smile\frown \rangle + A^{-1}\langle)(\rangle$

2. $\langle \bigcirc K \rangle = d\langle K \rangle$

   with $d = -A^2 - A^{-2}$

3. $\langle O \rangle = d.$

A Yang-Baxter model for the bracket is obtained as follows.

Choose a height function for the diagram, and divide the diagram into cups, caps and crossings. Delineate each of these critical points by placing **nodes** on the diagram as shown below:

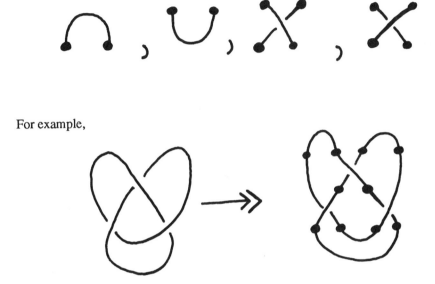

For example,

A **state** *(note that the states to follow are of a somewhat different character from the general combinatorial states for the bracket)* of the diagram is a choice of assignments of the indices {1,2} to the nodes of the diagram. With a given state, associate matrix elements to each maxima, minima or crossing as indicated below

$$(M_{ab}) = (M^{ab}) = M$$

$$M = \begin{bmatrix} 0 & \sqrt{-1}\,A \\ -\sqrt{-1}\,A^{-1} & 0 \end{bmatrix}$$

Here the lower and upper index entries of M denote the same elements of M - its ab entry, and the matrices R and $\overline{R}$ are defined by the formulas

$$R^{ab}_{cd} = A\,M^{ab}M_{cd} + A^{-1}\,\delta^a_c\,\delta^b_d$$

$$\overline{R}^{ab}_{cd} = A^{-1}\,M^{ab}M_{cd} + A\,\delta^a_c\,\delta^b_d$$

The Kronecker delta corresponds to an arc that is free of critical points with respect to this height function. Note that the matrix M is its own inverse so that the cancellations are automatic. Also the loop value is just right:

$$a\bigcirc b = \sum_{a,b}' M_{ab} M^{ab} = -A^2 - A^{-2}.$$

To each state   S: **Nodes**(K) -------> {1,2}, let <K|S> denote the product of the matrix entries that are produced by S at the critical points of the diagram.   Then <K> = Σ<K|S>   where the summation is taken over all possible states.   It follws at once that

$$\left\langle \,\slashoverback\, \right\rangle = A \left\langle \,\asymp\, \right\rangle + A^{-1} \left\langle \,)(\, \right\rangle$$

from the defining formulas for the R-matrices.   Since the loop value is correct, it is easy to see from the definition of the matrices that   <K> satisfies the identities 1., 2., 3.   above.   Abstract properties of the bracket make it easy to deduce that R satisfies the Yang-Baxter equation (see [110] and [114] for a discussion of this point.).

Example.

$$c\,\left(\!\!\begin{array}{c} a \\ d \end{array} \begin{array}{c} b \\ e \end{array}\!\!\right) f = \sum_{a,b,c,d,e,f,g,h}' M_{ca} M_{bf} R^{ab}_{de} R^{de}_{gh} M^{cg} M^{hf}.$$

Special value   A=-1 and the Penrose   Binor Calculus

With this model of the bracket in front of us, note how it behaves at the special value A=-1. Here we have

0. $<$$> = <$$> = $ (by def) $<$$>$.

1. $<$$> + <$$> + <$$> = 0.$

2. $<O> = -2.$

3. $<$$> = <$$> = <$$>$

In this special case, the matrix $M = \sqrt{-1}\ \varepsilon$ where ε is the matrix

$$\varepsilon \ = \begin{pmatrix} 0 & 1 \\ -1 & 0 \end{pmatrix}$$

This matrix , epsilon, is the defining invariant for the group SL(2) (read SL(2,C)). That is, SL(2) is the set of matrices of determinant 1, and epsilon has the property that for any matrix with commuting entries

$$P\varepsilon P^T = DET(P)\varepsilon.$$

In this context, we can interpret the equation 1. above as a matrix identity where the crossed arcs are Kronecker deltas:

$$\varepsilon^{ab}\varepsilon_{cd} = \delta^a_c \delta^b_d - \delta^a_d \delta^b_c$$

The network calculus associated with the value A=-1 corresponds to a diagrammatic calculus for SL(2) invariant tensors. This calculus was studied by Roger Penrose [166], [167] in order to investigate the foundations of spin, angular momentum and the structure of space-time.

Comment on the Yang-Baxter Equation and Rapidity

We can interpret the Yang-Baxter Equation as stating an identity about the scattering matrices for idealized two-particle interactions. But so far, the only properties that these particles had was a discrete quantity that we called "spin". The spin came from a discrete ordered index set.

To be more realistic, one can include a measure of relative momentum in the form of a *rapidity* parameter theta , θ . That is, the scattering matrix is a function of an angular parameter θ expressed in radians --

$$S = S^{ab}_{cd}(\theta).$$

A word of background [225] about this rapidity parameter is in order. The underlying assumption is that the particles are interacting in the context of special relativity. This means that the momenta of two interacting particles can be shifted by a Lorentz Transformation and the resulting scattering amplitude must remain invariant. Now , in the relativistic context (with light speed equal to unity) we have the fundamental relationship among energy (E), momentum (p) and mass (m):

$$E^2 - p^2 = m^2$$

(Note that this is in 1+1 dimensional spacetime.) Letting the mass be unity, we have

$$E^2 - p^2 = 1$$

and can represent $E = \cosh(\theta)$, $p = \sinh(\theta)$.

The Lorentz transformation shifts theta by a fixed amount. In light cone coordinates the Energy-Momentum has the form $[e^\theta, e^{-\theta}]$, and this is transformed via

$$[A, B] \text{-----}> [Ae^\phi, Be^{-\phi}]$$

under a Lorentz transformation. Consequently, if the two particles have respective **rapidities** (the theta parameter) θ_1 and θ_2, then the difference $\theta_1 - \theta_2$ is Lorentz invariant. The S-matrix must be a function of this difference of rapidities.

{A word about light cone coordinates: These are the coordinates $[t+x, t-x]$ where t and x denote one dimensional time and space coordinates respectively. With light speed equal to unity, the quantity $t^2 - x^2$ is invariant under change of inertial reference frame. Since $t^2 - x^2 = (t+x)(t-x)$, we see that the transform $[t+x, t-x] \text{-----}> [K(t+x), K^{-1}(t-x)]$ preserves $t^2 - x^2$, and hence represents a Lorentz transformation.}

Now consider the diagram in Figure 4. Since momentum is conserved in the interactions, we see that for the oriented triangle interaction, the middle term has a rapidity difference that is the sum of the rapidity differences of the other two terms. But the angles between the corresponding lines satisfy just this relation - by a Theorem well-known to Euclid!

Thus we can identify the rapidity differences with the angles between the interaction lines and obtain the correct angular relationships in the geometry of the diagrams.

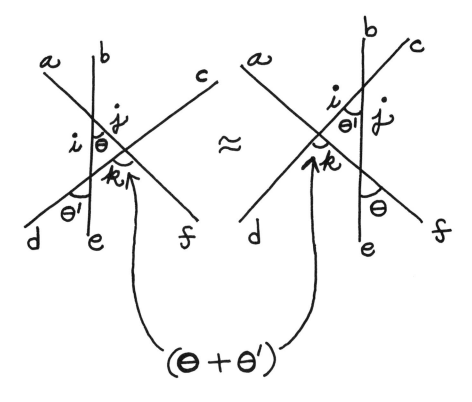

$$S_{ij}^{ab}\ (\theta)S_{kf}^{jc}\ (\theta+\theta')S_{de}^{ik}\ (\theta') = S_{ij}^{bc}\ (\theta')S_{dk}^{ai}\ (\theta'+\theta)S_{ef}^{kj}\ (\theta)$$

Quantum Yang-Baxter Equation with Rapidity

Figure 4

In all the physical applications the rapidity parameter is of great importance. For the sake of knot theory and quantum groups it is usually ignored (by taking a limiting case, or by constructing solutions that do not depend upon an extra parameter of this sort). Of course, if one does not have this extra dependence, then the Yang-Baxter solution immediately gives rise to a representation of the Artin braid group.

Nevertheless it is an good question to ask if the rapidity can be used in topology. In [110] it is shown how use of the rapidity can be used to construct rapidity-free solutions in transfering between solutions coresponding to SL(n) (Homfly polynomial)   and   SO(n)   (Dubrovnik polynomial).   Recently, Vaughan Jones (Lecture at MSRI - March 1991) has found a use for the rapidity in the case of the V-polynomial, by identifying $S(\theta)$ as a certain tangle (rather than just a crossing). This allows the transfer of a basic argument in statistical mechanics to the knot theory. The result is a new method for producing collections of links that have the same Jones polynomial - hence a better understanding of the limitations of this topological invariant.

The Quantum Group   SL(2)q

The calculus of link evaluations via the bracket polynomial provides a significant generalization of the Penrose spin networks. This is obtained by shifting from SL(2) at A=-1 to SL(2)q at $q = \sqrt{A}$ for arbitrary A, where SL(2)q denotes the quantum group in the sense of Drinfeld. Here is how SL(2)q arises in this context:

Since SL(2) is characterized by matrices P (with commutative entries) such that $P \varepsilon P^T = \varepsilon,$ and since the more general matrix

$$\varepsilon^\wedge = \begin{pmatrix} 0 & A \\ -A^{-1} & 0 \end{pmatrix}$$

is the basis for constructing the bracket polynomial, it is natural to ask what sorts of matrices will satisfy the equations of invariance with ε replaced by $\varepsilon^\wedge$:

(*) $P \varepsilon^\wedge P^T = \varepsilon^\wedge$

$P^T \varepsilon^\wedge P = \varepsilon^\wedge$

Attempting to generalize invariance in this way leads to well-known difficulties. It is necessary to assume that the entries of P do not necessarily commute with one another. Assume that P has the form

$$P = \begin{pmatrix} a & b \\ c & d \end{pmatrix}$$

where a,b,c,d belong to an associative (not necessarily commutative) ring. It is then an exercise in elementary algebra to see that the equations (*) are equivalent to the system of relations shown below.

$$ba=qab \qquad ca=qac$$
$$dc=qcd \qquad db=qbd$$
$$bc=cb$$
$$ad - da = (q^{-1} - q)bc$$
$$ad - q^{-1}bc = 1$$

where $q = \sqrt{A}$.

The entries of the matrix P form a non-commutative algebra, $\mathbb{A}$.

The algebra $\mathbb{A}$ is a Hopf algebra with coproduct

$$\Delta: \mathbb{A} \dashrightarrow \mathbb{A} \otimes \mathbb{A}$$

given by the formula

$$\Delta(a) = a \otimes a + b \otimes c$$
$$\Delta(b) = a \otimes b + b \otimes d$$
$$\Delta(c) = c \otimes a + d \otimes c$$
$$\Delta(d) = c \otimes b + d \otimes d.$$

The antipode is determined by the fact that P is invertible with respect to the algebra of its own elements. We have $s: \mathbb{A} \dashrightarrow \mathbb{A}$, the antipode with

$$s \begin{pmatrix} a & b \\ c & d \end{pmatrix} = \begin{pmatrix} d & -qb \\ -q^{-1}c & a \end{pmatrix} .$$

This Hopf algebra is the quantum group $SL(2)q$. Thus $SL(2)q$ arises quite naturally from the bracket model of the Jones polynomial.

Antipodal Remark. There is not room in this paper to go into the remarkable fit between Hopf algebras and link invariants. However, it is worth remarking that in the category of morphisms explained at the beginning of this section there is a natural map that acts as an anti-morphism of $End(V)$. This map is a composition of cap, transpose and cup as shown below: $S(a) = Cap \, a^T \, Cup.$

$$S(ab) = S(b) S(a).$$

This is the beginning of the general development of the relationship with Hopf algebras. See [114], [104],[65], [66], [148], [177], [221]. Of particular interest are results of Hennings and Reshetikhin that show that the existence of 3-manifold invariants via surgery and framed links is intimately tied to the existence of integrals for finite dimensional Hopf algebras. There are genuinely fundamental relationships between the theory of Hopf algebras as quantum groups and the structure of these invariants. Papers [134], [104] and [105] mark the beginning of an investigation into the roots of the relationship of quantum groups and combinatorial knot theory. In [104] it is showm how the universality of the form of this invariant is related to properties of the Gauss code for diagrams. In [105] ribbon elements in Drinfeld doubles are algebraically characterized.

Spin Networks Remembered

Classical Penrose spin networks are based on the binor calculus (see below), and they are designed to facilitate calculations about angular momentum and SL(2). A spinor is a vector in two complex variables, denoted by Ψ^A, A=1,2. The spinor space is acted on by elements U in SL(2) so that

$$(U \, \Psi)^A \;=\; \Sigma U^A{}_B \Psi^B$$

(Einstein summation convention). A natural SL(2) invariant inner product on spinors is given by the formula $\Psi\Psi^*$ where

$$\Psi^* = \varepsilon_{AB}\Psi^B$$

so that $\quad \Psi \, \Psi^* \;=\; \Psi^A \varepsilon_{AB}\Psi^B \quad$ (sum on A and B). If we

wish to diagram this inner product, then we let

It is natural to lower the index via

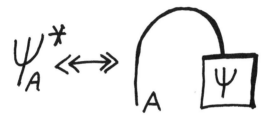

Then

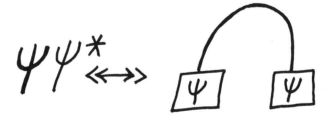

and the fragment ⌢ is interpreted as the epsilon, ε_{AB}.

Penrose made special conventions for maxima or minima (a minus sign for the minima) in order to insure planar topological invariance.

$$\cap \langle\!\langle\leftrightarrow\rangle\!\rangle \; \varepsilon \; , \; \cup \langle\!\langle\leftrightarrow\rangle\!\rangle -\varepsilon$$

$$(\text{Penrose Convention})$$

These conventions are equivalent to choosing to replace ε by $\sqrt{-1}\,\varepsilon$. Thus in the classical spin nets

$$\int_{a}^{c}{}_{b} = \int_{a}^{c} \langle\!\langle\leftrightarrow\rangle\!\rangle (\sqrt{-1}\,\varepsilon_{ab})(\sqrt{-1}\,\varepsilon^{bc})$$

$$= \delta_a^c$$

and

$$\times^{a\;b}_{c\;d} \langle\!\langle\leftrightarrow\rangle\!\rangle -\delta_d^a \delta_c^b$$

Note that the loop value is

$$\bigcirc = (\sqrt{-1})^2 + (\sqrt{-1})^2 = -2.$$

This calculus entails the binor identity

$$\bigtimes + \smile\!\!\frown +)(= \emptyset$$

and   (directly or from the fact that we are looking at a special case of the bracket) it is invariant under the projections of the Reidemeister moves. Therefore, any loop, even with self-crossings, has value -2 . The binor calculus is a unique planar calculus associated with both the bracket polynomial and the representations of SL(2).

The next important spin network ingredient is the **antisymmetrizer.** This is a diagram sum associated to a bundle of lines, and is denoted by

$$\boxed{\bigtimes\bigtimes\bigtimes}\,^{N}$$

where the   N   denotes a bundle of parallel strands of multiplicity N.

The antisymmetrizer is defined by the formula

$$\boxed{\bigtimes}\,^{N} = \frac{1}{N!} \sum_{\sigma \in S_N} \text{sgn}(\sigma)\, \boxed{\sigma}$$

where σ runs over all permutations in S_n, $\text{sgn}(\sigma)$ is the sign of the permutation, and the σ in the box denotes the diagrammatic representation of this permutation as a braid projection.

As discussed in the previous section, this entire context generalizes to the q-deformed spin networks by using the bracket expansion in place of the binor identity. It should be clear from the this section that the generalization entails precisely the replacement of SL(2) by the quantum group SL(2)$_q$.

Remarks. With this widened view of the spin network theory it is possible to ask more questions. The idea of using a partition function on a three dimensional manifold via spin network evaluations goes back to work of Regge, Ponzano, Hasslacher and Perry ([172],[61],[159]). (See recent remarks on this context in [34],[107].) The motivation for these earlier works was a parallel between the formalism of functional integral quantization of 2+1 quantum gravity (2 space dimensions and 1 time dimension) and the semi-classical limit of the partition function defined on a three-dimensional space whose boundary was the space for this mathematical space-time. The present reformulation of spin networks over SL(2)$_q$ calls all this prior work out for re-examination. We know, that by working with q a root of unity that it is possible to obtain well-defined finite partition functions (the Turaev-Viro invariants) that are a significant construction in this area of three-dimensional topology. It is possible that these same partition functions will play a useful role in quantum gravity. In any case, the context of quantum gravity gives a new way to look at the topological invariants and may lead to the construction of significant examples of three dimensional manifolds.

We need new sources of intuition about the behaviour of these partition functions on closed 3-manifolds. It is possible that there exists a compact 3-manifold with non-trivial Turaev-Viro invariant, but trivial fundamental group (giving a counterexample to the Poincare conjecture), but the nature of such an example has long eluded the conceptual frameworks of three dimensional topologists. I believe that the more closely we look into the mathematical physical structure of 3-manifolds, the better is the chance of finding such an example.

Beyond $SL(2)_q$.

This discussion has concentrated on the SL(2) quantum group. A next stage for the interlock of combinatorics and quantum groups in the domain of 3-manifold invariants and recoupling theory is the the SL(n) quantum group. The representation theory of this group is considerably more complicated, but there is an analog of the bracket polynomial for this case (See [110],[107]). This analog is a generalization of a Yang-Baxter model of Jones [81] with the following properties:

A state consists in an oriented splitting 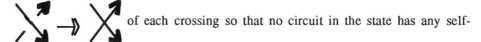 or a projection

of each crossing so that no circuit in the state has any self-crossings. Each circuit is labelled with a spin from the given index set. I use the following notation for spin relations at a projected crossing or at a split :

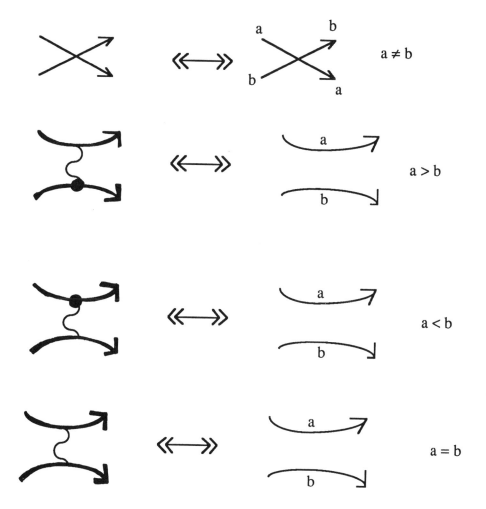

The same forms can be used as generalized "Kronecker deltas". Thus :

$$
\left.\substack{a \qquad b \\ \\ \\ c \qquad d}\right\} = \begin{cases} 1 & \text{if } a = c \ < b = d \\ \\ 0 & \text{otherwise} \end{cases}
$$

The model then has the expansion as a generalized bracket :

$$\left(Z = W - W^{-1}\right)$$

$$\left[\,\diagdown\!\!\!\diagup\,\right] = Z \left[\,\smile\!\!\!\frown\,\right] + W \left[\,\rlap{=}\,\right] + \left[\,\diagdown\!\!\!\diagup\,\right]$$

$$\left[\,\diagdown\!\!\!\diagup\,\right] = -Z \left[\,\wedge\!\!\!\vee\,\right] + W^{-1} \left[\,\rlap{=}\,\right] + \left[\,\diagdown\!\!\!\diagup\,\right]$$

Such an expansion is equivalent to giving the vertex weights. In particular, the Yang-Baxter tensor for this model is given by

$$R^{ab}_{cd} = Z \quad + W \quad +$$

$$\overline{R}^{ab}_{cd} = -Z \quad + W^{-1} \quad +$$

Note that since

$$\left[\,\smile\!\!\!\frown\,\right] = \left[\,\vee\,\right] + \left[\,\wedge\,\right] + \left[\,\rlap{=}\,\right]$$

is a tautology, we have the exchange identity

$$\left[\ \searrow\hspace{-0.3em}\nearrow\ \right] - \left[\ \searrow\hspace{-0.3em}\searrow\ \right]\ =\ Z\left[\ \rightleftharpoons\ \right]$$

when $Z = W - W^{-1}$.

The global evaluation of a state is given by the formula

$$\|S\| = \sum_{C \text{ a circuit in } S} \text{label}\,(C) \cdot \text{rot}\,(C)$$

where

$$\text{rot}\!\left(\ \circlearrowleft\ \right) = +\,1$$

$$\text{rot}\!\left(\ \circlearrowright\ \right) = -\,1$$

Thus

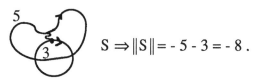

$$S \Rightarrow \|S\| = -\,5 - 3 = -\,8\ .$$

This global evaluation is a direct generalization of the counting of circuits in the bracket. The rotation numbers rot(C) are special cases of the Whitney degree of a plane curve.

It is an interesting exercise to verify directly that this model satisfies the Yang-Baxter equation (invariance under oriented type III move). One finds that this invariance occurs for *any ordered index set*. For any finite ordered index set regular isotopy invariance then becomes the demand that

$$W^{-1}\left[\ \bigcirc\ \right] = W\left[\ \bigcirc\ \right]$$

and

$$W^{-1}\left[\ \bigcirc\ \right] = W\left[\ \bigcirc\ \right]$$

If we require multiplicativity,

$$\left(\left[\ \bigcirc\rightarrow\ \right] = a\left[\ \rightarrow\ \right], \left[\ \bigcirc\rightarrow\ \right] = a^{-1}\left[\ \rightarrow\ \right]\right),$$

then everything simplifies. The model imposes a very strong requirement on the index set, and I prove that the index set must be of the form {-n,-n+2,...,n-2,n} . This is exactly the spectrum of spins that Jones used in his model [81] coming from the SL(n) quantum groups.

In fact, one can construe this model, without the indices, as a specific graphical expansion for the Homfly polynomial that is consistent (since the SL(n) models exist) and dependent on specific graph evaluations (such as those derived from the index decorations). The model, at this level of abstraction, encodes combinatorial properties of the Hecke algebra and of the SL(n) quantum group. It is an interesting and intricate problem to investigate the analogues of the Jones projectors and the recoupling theory using this model. This is work in progress that should yield combinatorial insight for the three manifold invariants associated with $SL(n)_q$.

IV. Yang-Baxter Models and the Alexander Polynomial

The Free Differential Calculus [35] of Ralph Fox gives a method for defining the multi-variable Alexander polynomial for a colored link (each component has its own label) in terms of a presentation of the fundamental group of the complement of the link. In [101] and [102] we show how to obtain solutions of the Yang-Baxter equation and corresponding models for the multi-variable Alexander polynomial by starting from representations derived from the free differential calculus. Thus we manage to dig Yang-Baxter solutions out of the fundamental group of the link complement. In the process we relate these constructions to the so-called free fermion model in statistical mechanics, and derive a new determinant formulation for the multivariable Alexander polynomial. This new determinant arises from the use of Grassmann variables in the free fermion model.

These constructions related to the Alexander polynomial give rise to many questions. First of all, the Alexander polynomial was always pleasant to work with because of its multiplicity of interpretations. One could regard it as a determinant of a Jacobian matrix in the free calculus, or as a generator of the annihilator ideal for the first homology of the infinite cyclic cover of the link complement (as a module over $Z[t,t^{-1}]$ where t generates the covering transformations of this cover), as the determinant of a matrix derived from the Seifert pairing for a spanning surface for the link in three dimensional space, or as the result of a skein calculation using the Conway formulation. The geometric side of these interpretations seems to disappear for the Jones polynomial and its generalizations.

One application of this statistical mechanics approach to the Alexander polynomial is Murakami's recent work [162] on new skein relations for the multi-variable polynomial. His work gives implicitly a skein algorithm for computing the multi-variable Alexander polynomial. It is an excellent challenge to find a simpler version of his result - in particular to find a simple proof that his extra identities are sufficient for skein computation.

One consequence of this disappearance is the lack of analogues for certain theorems about the Alexander polynomial. In particular, it is a well-known fact due to Fox and Milnor [45] that the Alexander polynomial of a slice knot K has the form $f(t)f(t^{-1})$, where $f(t)$ is a polynomial in t. A slice knot is a knot that bounds a smooth disk in the four dimensional ball whose boundary is the ambient three sphere for the knot in question. *Is there an analogue of the Fox-Milnor theorem for the Jones polynomial?*

So far nothing is known about this question. In the context of this report, there are three lines of approach to the problem. *Line 1* : Stick to ribbon knots (a combinatorially defined class of slice knots that is conjectured to exhaust the category of slice knots) and **transfer a calculational proof of the Fox-Milnor theorem for ribbons from classical determinants to state summations via our work on the free-fermion model.** The outcome of this line will be a viewpoint on the Fox-Milnor Theorem from a Yang-Baxter model for the Alexander polynomial. With this viewpoint in hand the next stage is to look for a comparison with the Yang-Baxter model for the Jones polynomial. Since these models are very similar in structure, this program may show the way to the analogue theorem.

Line 2 : As we have remarked earlier, the Reidemeister torsion of a 3-manifold appears in averaged form in the Witten version of the Reshetikhin-Turaev invariant. In the original form of the Fox-Milnor Theorem, the Alexander polynomial is interpreted as a Reidemeister torsion. Thus the Witten invariant should hold a message about the

Alexander polynomial if we could decode it. Line 2 has been observed by many people. Our opinion is that Line 1 will give information for Line 2 as well. *Line 3* : Look at four dimensional formulations of the Witten approach to link invariants.

Any information about the behaviour of the generalized polynomials on ribbon knots or slice knots will be very welcome, since this arena is just on the border between three and four dimensions, and because
the detection of slice knots is a significant problem in classical knot theory. Also, this problem borders on the question of the construction of invariants of knotted surfaces in four-space in some way that generalizes the partition function constructions that work for the Yang-Baxter models in three space. Recent work of Carter and Saito [25] on generalized Reidemeister moves for surfaces in four-space indicates that such generalizations may exist.

To return to the original statement of this section. It is possible to dig certain Yang-Baxter solutions from the classical topology in relation to the Free Differential Calculus. Here is a sketch of this method. The operators in the free calculus are special derivations defined on the group ring $Z[G]$ of a group G. If a and b are words in G, then such a derivation D satisfies the rule

$$D(ab) \ = \ D(a) \ + \ bD(b).$$

The basic motivation for this rule comes from the equation

$$l(ab) \ = \ l(a) \ + \ al(b)$$

where $l(x)$ denotes the lift of an element in the fundamental group G of a space X into the universal covering space $U(X)$. Thus $l(x)$ is a path starting at the basepoint of $U(X)$. The group G acts as the group of automorphism of the discrete bundle

U(X) over the base space X. The term al(b) denotes the result of acting on l(b) by the automorphism corresponding to a. Thus the initial point of al(b) is the end point of l(a), and the equation expresses a decomposition of this path. In the free calculus one obtains operators for each element in the basis of the presentation of G. This derives from the fact that one can decompose a path lift in the universal covering space of a free group into a sum of products of elements of the group ring of the group multiplied by basic edges that are in one-to-one correspondence with the group itself.

In any case, a Fox derivation is easily seen to satisfy the property

$$D(a\text{-}1) = -a^{-1}D(a)$$

(since $0 = D(1) = D(aa\text{-}1) = D(a) + aD(a^{-1})$). Consequently,

$$D(bab^{-1}) = D(b) + b[D(a) + a[-b^{-1}D(b)]] = bD(a) + (1 - ab^{-1})D(b).$$

We rewrite the form of this equation as

$$D(c) = bD(a) + (1 - ab^{-1})D(b).$$

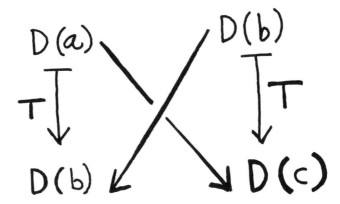

(In the Wirtinger presentation [35] of the fundamental group one has the relation $c = bab^{-1}$ at a crossing in the diagram. In this presentation, the fundamental group is generated by one loop for each arc in the diagram.)

One then obtains the colored Burau representations of the Artin braid group by identifying $D(a)$, $D(b)$ and $D(c)$ as vector space generators corresponding to the arcs on a diagram (an arc goes from one crossing to another) at a crossing as shown above. The small letters a,b,c are regarded as *commuting* polynomial variables associated with the components of the diagram. The linear transformation T on the two dimensional vector space associated with the crossing is given by

$$T(D(b)) = D(c) = bD(a) + (1 - ab^{-1})D(b).$$

$$T(D(a)) = D(b)$$

The Yang-Baxter solutions are obtained by looking at the action of this representation on the exterior algebra of the original module V. Let $T^{\wedge}$ denote the extension of T to the exterior algebra. Let $v=D(a)$, $w=D(b)$. Then $\{v,w\}$ is a basis for the module V, and $L^*(V)$ has basis $\{1,v,w,v \wedge w\}$ where $\wedge$ denotes the exterior product. We have

$$T^{\wedge}(1)=1$$
$$T^{\wedge}(v) = w$$
$$T^{\wedge}(w) = bv + (1 - ab^{-1})w$$
$$T^{\wedge}(v \wedge w) = Det(T)v \wedge w = bv \wedge w$$

The linear map $T^{\wedge} : L^*(V) -----> L^*(V)$ satisfies the Yang-Baxter equation.

This procedure leading from the fundamental group through the Burau representation to solutions to the Yang-Baxter equation leads directly to the problem: *Can the machinery of the Free Differential Calculus be generalized so that other Yang-Baxter solutions are generated from it?*

This is part of the more general problem: *Give a construction of the Jones polynomial and its generalizations that derives from the structure of the fundamental group of he link complement and peripheral subgroups (the subgroups of loops that are represented on the surface of a tubular neighborhood of the link).*

It is a fun to reformulate and explore the original context of the Free Differential Calculus in search of clues for generalizations. In this context there is a theme of non-commutative geometry that has its modern analogues ([26],[149]).

There should be relationships between these questions and the properties of the Vassiliev invariants discussed in the next section.

The Vassiliev invariants return to classical topology by considering the structure of combinatorial knot space.

V. Vassiliev Invariants - Underlying Combinatorics

By examining the topology of the space of embeddings of a circle into three dimensional space Vassiliev [209] has discovered a new context for producing invariants of knots and links. Birman and Lin [19] have shown that the Vassiliev invariants produce the Jones polynomial and some of its generalizations. In particular, Birman and Lin show that If $V_K(t)$ is the original Jones polynomial and if $v_i(x)$ denotes the coefficient of x^i in $V_K(\exp(x))$, then v_i is a Vassiliev invariant of order i. This means that vi satisfies a switching identity of the form $v_i(\times) - v_i(\times) = v_i(\times)$ where $v_i(\times)$ denotes a rigid vertex graph invariant that vanishes for graphs with more than i vertices. The graph invariants at the basis of the Vassiliev scheme are determined by a spectral sequence related to the space of knot embeddings.

It is of interest to examine the combinatorics of the Vassiliev invariants both by the Vassiliev methods, and by what we already know about existing knot polynomials. In the case of the Jones polynomial, one can use the spanning tree expansion for the Jones polynomial ([198],[96], [92]) to obtain explicit expressions for these Vassiliev invariants. This method will be indicated here in the context of the bracket polynomial. Here the spanning tree expansion can be expressed in terms of Jordan-Euler trails on the knot diagram [87]. A Jordan-Euler trail (**trail** for short) is a walk along the edges of the diagram that goes through every edge once, and declines to cross at every crossing. In terms of bracket states, a trail is simply a state with one component. Thus the trails for the trefoil diagram are as shown below.

The bracket can be calculated from the trails by a prescription that involves a notion of *activity* for each site in the trail. A **site** consists of two nearby arcs from the trail whose recombination produces a crossing from the original knot diagram. Label each crossing by a distinct positive integer. Each site in a trail receives a corresponding label. **Activity** is defined as follows: Take a site in a trail T. Resmooth it in the opposite manner. This causes the trail to decompose into two Jordan curves. These two curves meet each other at a subset of the sites including the site in question. Each of these sites has a label. Call the original site **active** if its label is least among this subset of sites. The diagrams above indicate the activities for the trails on the trefoil diagram (with respect to a given labelling).

Let $i_+(K,T)$ denote the number of inactive sites that are A-smoothings of K. Let $i_-(K,T)$ denote the number of inactive sites that are B-smoothings of K. (The terminology A and B smoothings is defined in section 1.) Let $a_+(K,T)$ denote the number of active B-smoothings and let $a_-(K,T)$ denote the number of active A-smoothings. Let $r(K,T) = i_+(K,T) - i_-(K,T) + 3a_+(K,T) - 3a_-(K,T)$ and let $a(K,T) = a_+(K,T) + a_-(K,T)$. Then $<K>$ (topological bracket) is given by the following formula

$$<K> = \Sigma_T \; A^{r(K,T)} \; (-1)^{a(T)}$$

where T runs over all Jordan Euler trails on the diagram K.

This formula is useful in its own right, particularly for considering the question whether the bracket can be used to detect knottedness.

If the substitution $A = \exp(x)$ is performed, then we obtain

$$<K>(\exp(x)) = \Sigma_T \; (\exp(x))^{r(K,T)} \; (-1)^{a(T)}$$

$$= \Sigma_T \; (\exp(xr(K,T)) \; (-1)^{a(T)}$$

$$= \sum_T \sum_{n=0}^{\infty} (1/n!)x^n r(K,T)^n (-1)^{a(K,T)}$$

$$= \sum_{n=0}^{\infty} x^n (1/n!) \sum_T r(K,T)^n (-1)^{a(K,T)}$$

Thus the coefficient , $C(n)$, of x^n in this expansion of the bracket polynomial is given by the formula

$$C(n) = (1/n!)\sum_{T} r(K,T)^n (-1)^{a(K,T)} \ .$$

This coefficient is the analogue, for the bracket, of the n-th Vassiliev invariant. By the same method, we can obtain a specific formula for the n-th Vassiliev invariant itself via the formula relating Jones polynomial and bracket:

$$V_K(t) = [(-A^3)^{-w(K)} <K>(A)](A = t^{-1/4}).$$

As a result, there is an elegant formula for the Vassiliev invariants related to the Jones polynomial in terms of Tutte activities in a spanning tree expansion (via the trail representation of the trees).

A deeper picture of this relationship will emerge as these formulas are seen to come directly from the Vassiliev scheme. This will mean the beginning of new interpretations of the combinatorial state models in terms of the topology of the space of knot embeddings.

VI. Integral Heuristics

Recall that the Homfly polynomial may be expressed in regular isotopy form via a skein identity and a specification of behaviour under a "twist", or type I Reidemeister move:

1.SKEIN IDENTITY

2.TWIST BEHAVIOUR

$$H\ \sigma = a\ H , \quad H\ \sigma = \bar{a}\ H\ .$$

It is assumed here that the polynomial H is invariant under the type two and type three Reidemeister moves. One then obtains the usual Homfly polynomial, P_K , by defining $P_K = a^{-w(K)}H_K$ where w(K) denotes the twist number of K. The polynomial P is then an invariant of ambient isotopy for oriented links K.

Another way to think about this definition is to regard K as a framed link. This means that K is endowed with a normal vector field, or equivalently that K is replaced by an embedding of a collection of bands, one band per link component. H is then an ambient isotopy invariant of the embedding of the bands.

Now suppose that W is a functional on knots and links that is represented by an integral formalism

$$W_K = \int dA\ e^L\ T_K(A)$$

where we assume that **T** is not a topological invariant of K , but that **T** satisfies a local skein identity of the form

$$T\ -\ T\ \ \ = -zF\ T^*\ \ \ .$$

Here **F** is a function that depends upon the point where the strands are switched (the exact dependence is un-specified).

On the right-hand side of this skein identity we have written **T*** rather than **T**. We shall assume [116] that **T*** satisfies the equation

$$\delta T^*{}_K/\delta A = T_K.$$

We assume that **L** satisfies the relationship

$$\delta L/\delta A = F.$$

Finally, we assume that in an integration by parts involving the integral

$$\int dA \; e^L \; T_K(A)$$

the boundary terms vanish.

Then the skein relation for **W**$_K$ is a formal consequence:

Proposition: $\quad W_{\text{⤳}} - W_{\text{⤳}} = z W_{\text{→}}$.

Proof. $\quad W_{\text{⤳}} - W_{\text{⤳}} = \int dA\, e^L T_{\text{⤳}} - \int dA\, e^L T_{\text{⤳}}$

$$= \int dA\, e^L [T_{\text{⤳}} - T_{\text{⤳}}]$$

$$= \int dA\, e^L (-zFT^*{}_{\text{→}})$$

$$= -z \int dA(e^L F)T* \,\rightrightarrows$$

$$= -z \int dA(\delta e^L/\delta A)T* \,\rightrightarrows$$

$$= z \int dA e L(\delta T* \rightrightarrows /\delta A)$$

$$= z \int dA e^L T \,\rightrightarrows$$

$$= zW \rightrightarrows \qquad\qquad //$$

If we assume that $T_{\,\curvearrowright} = -aFT*_{\rightarrow}$ and $T_{\,\curvearrowleft} = -a^{-1}FT*_{\rightarrow}$
then it follows by the same formalism that

$$W_{\,\curvearrowright} = aW_{\rightarrow} \qquad \text{and} \qquad W_{\,\curvearrowleft} = a^{-1}W_{\rightarrow} \quad.$$

Discussion

Compare the formalism of our heuristic integral with the Witten
functional integral [215] (recall the discussion in the introduction to these notes).
Then L is an analogue of the integral of the Chern-Simons Lagrangian, and T is an
analogue of the trace of the holonomy of a gauge potential A along K. It is a fact
that the variation of the integral of the Chern-Simons Lagrangian with respect to the
gauge A (taken at a point p) gives the curvature tensor F for the gauge potential taken
at that point. Furthermore, in the gauge field theory of the Witten integral the
curvature at a point in space is measured by examining the holonomy of A around a
very small curve enclosing this point. The two ways of measuring curvature interact to
make this integral work. It is this interaction that is represented in schematic form in
the heuristic described here.

Note that the difference in holonomy of

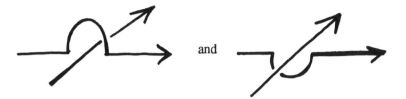

and

will involve the holonomy around a small loop surrounding the switch point --picking up the curvature at that switch point geometrically. Thus the equation

$$T \cdot T = -zFT^*$$

can be regarded as an expression of this dependence of change of holonomy on the local curvature. Finally the presence of T^* with

$$\delta T^*/\delta A = T$$

denotes a local compensation so that the coupling factor z is a constant. With these stipulations, we see that the heuristic fits the *context* of the Witten integral. The literal details are different, but the theme of integration by parts and the two interpretations of curvature carry over.

To go further involves using the full formalism of the functional integral -- including a representation of the gauge group and a direct treatment of the effect of strand switching on the holonomy. This calculation has been initiated in [191] and continued in [116], [31], [107]. In the full formalism the story is similar, and one sees how the properties of the Lie algebra of the gauge group conspire to give the skein relation to a first order of approximation.

I have introduced this integral heuristic to indicate a realm that deserves deeper investigation. This is the realm of formal correspondences that go part and parcel with the workings of functional integral formalisms at the physical level of rigor. There are many enticing partial results in this domain, some of the most interesting being those obtained by the Italian group [58], and in the papers of Dror Bar-Natan [13] , [14] and in recent work of Singer and Axelrod [9]. In some cases terms from the functional integral can be proved independently to yield invariants. In the case of the Italian work, there is an explicit (classical) integral formula for the second coefficient of the Conway polynomial and a hint of generalizations to higher order terms. The point that is made with the heuristic, and particularly with its more serious counterpart using the gauge theory, is that there is a strong component of these invariants that is carried by the formalism of these integrals and an associated differential algebra. The question is: *how to separate the truly combinatorial part of this structure.*

The question of existence of a measure theory for the gauge connections modulo gauge equivalence is certainly important, but for the combinatorial topology it is possible that all that is necessary is a new formalism that maintains the point of view, but works over a finite ground.

The Turaev-Viro invariant is a case in point. In sections 2 and 3 we have shown how this partition function on a three manifold can be understood on purely combinatorial grounds. If it were possible to transfer the manipulations of the functional integral itself to a finite basis, then the relations with Reidemeister torsion, orientation, and other matters would become transparent. In advocating a search for a finite foundation for functional integrals, I am not necessarily asking for something like lattice gauge theory. In fact the various constructions of topological quantum field theories using the duality structure of conformal field theories [33],[125] are precisely examples of the sort of discretization that is sought. The problem with these sorts of models is that they completely take leave of the beautiful structure of differential tensor analysis that is so intimately tied with the formalism of the functional integral. In this sense, all these models throw out the baby with the bath water. Our problem is to discover how to save this baby - the generalized Feynman path integral - by giving it a combinatorial home.

Appendix 1 - The Two-Variable Skein Polynomials

This section reviews the properties of two two-variable skein polynomials - the Homfly poynomial ([49],[139],[173]) and the Kauffman polynomial [93]. Each of these polynomials generalizes the original Jones polynomial [78]. The Homfly polynomial is a generalization of the Alexander-Conway polynomial as well. The Kauffman polynomial generalizes the one-variable unoriented polynomial of Millett, Brandt Lickorish [21] and Ho [67].

The Homfly polynomial , P_K, may be described via the equation

$$P_K = a^{-w(K)}H_K$$

where the polynomial H_K is an invariant of regular isotopy satisfying the equations

$$H \nearrow \quad -H \nearrow \quad =zH \rightrightarrows$$

$$H \circ\!\!\rightarrow \quad =aH \qquad ,H \multimap\!\!\rightarrow \quad =a^{-1}H$$

$$H \circlearrowleft \quad =1.$$

For a=1, the polynomial H_K coincides with the Alexander-Conway polynomial. The description of the Alexander polynomial in the form of an exchange identity is due to

Conway [30]. The Conway version normalizes the Alexander polynomial, and extracts more information from it in the process (For example, it can detect chirality for some links with an even number of components.

The Homfly polynomial, an ambient isotopy invariant, can be expressed via the more direct identity

$$(*) \qquad aP \qquad -a^{-1}P \qquad =zP$$

The original Jones polynomial, $V_K(t)$, corresponds to the substitution $a=t^{-1}$, $z =\sqrt{t} - 1/\sqrt{t}$:

$$t^{-1}V \qquad - \quad tV \qquad =(\sqrt{t} - 1/\sqrt{t})V$$

REMARK ON FRAMING

It is of interest to interpret the extra variable a in the Homfly polynomial as measuring framing for the link. This framing information can be regarded as an interpretation of the writhe , w(K). That is , replace the link diagram by thickening each link component into a "flat" band.

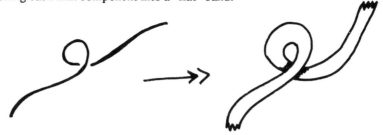

This flat band is obtained by drawing a second curve parallel to the original curve for each component and regarding the two parallel curves as forming an embedded band in three-space. Then for a knot, the writhe of the knot diagram (i.e. the sum of the signs of its self-crossings) is equal to the linking number of the two edges of the flat band. For a link , the writhe contributions between components of the link give twice the linking number of these components. Thus we have the formula

$$w(K) = \sum lk(\text{band boundaries}) + 2\sum lk(\text{band to band}).$$

From the point of view of the bands , this writhe , $w(K)$, is invariant topological information about the framing (A framing of the original link is equivalent to the specification of a band embedding.)

Since we regard P as having value 1 on the trivial knot, the equation (*) can be regarded as a mode of compensation so that the total framing is set to zero for each member of the skein triple

It should be mentioned that the term skein is also due to Conway , and refers to the collection of knots and links obtained from the given knot or link by switching and splicing crossings. The recursive calculation of the polynomial by switching and splicing is called a **skein calculation.**

It follows from the rules for these polynomials that

$$H_{0\,K} = d\; H_K\,, \qquad d = (a\text{-}a^{-1})/z$$

and similarly for P.

The Kauffman polynomial [88], [93], 107],

$$F_K[a,z],$$

is also defined as a writhe-normalized version of a regular isotopy invariant polynomial.

$$F_K = a^{-w(K)}L_K$$

Here the polynomial

$$L_K(a,z)$$

is a regular isotopy invariant for unoriented links , and satisfies the unoriented skein identities

In this case , the polynomial F cannot be described by a single skein identity, but rather by two such identities depending upon a connectivity condition. This makes the normalized regular isotopy description of the invariant simpler than any standard oriented skein version.

Appendix 2 - Knot Epistemology

In general, the structure of link diagrams involves a controlled mixing of levels that is not directly available at the usual set-theoretic ground floor of mathematics. To see this quite clearly, interpret a crossing as a "membership relation" -- a ε b :

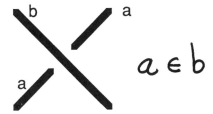

$$a \in b$$

The under-crossing arc is here taken to be a singly labelled entity, and "being under" is identified with membership. Linking gives rise to "sets" that are mutually related, and self-crossing to sets that are members of themselves.

This knot-set theory is directly related to the topology via the Reidemeister moves (See section 2 for a discussion of the moves). The second move tells us that this set theory is necessarily **fermionic** in the sense that **identicals vanish in pairs.**

$$b = \langle a, a, \ldots \rangle \rightsquigarrow b = \langle \cdots \rangle$$

(If knot-sets are fermionic, then **ordinary sets are *bosonic*,** in the sense that **identicals condense in pairs:** $\{1,1,1\}=\{1,1\}=\{1\}$.) Let $\langle x,y,z,\ldots \rangle$ denote a knot-set with members $x,y,z,\ldots$ Then we assume that $\langle a,a,b,\ldots \rangle = \langle b,\ldots \rangle$. Thus $\langle a,a,a \rangle = \langle a \rangle$.)

With this rule for dealing with identicals coupled with a free attitude toward self-membership (corresponding to the move of type 1) as shown below,

$$a = \langle a, \ldots \rangle \quad \rightleftharpoons \quad a = \langle \ldots \rangle$$

it is easy to see that *topologically equivalent diagrams give rise to the same knot-set. Note that in this version of knot-sets, self-membership is factored out by the type 1 move, but linking in the form a=<b> , b=<a> remains.*

In order to make a topological model of knot-sets that actually includes the self-membership, it is necessary to use *framed links*. A framed link is a link such that each component has a normal vector field. This is adequately modelled topologically by replacing each link component by an embedded band with some twists in it. A twist by 360 degrees is topologically equivalent to a curl, which we have here identified as a self-membership:

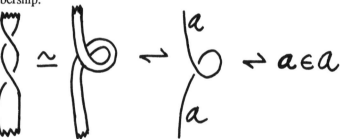

For elemental knot set theory one only wants to know whether or not a set belongs to itself. In the light of double cancellation of membership, as described above, we shall add the rule that twists are counted modulo 2π and that only multiples of 2π twists are allowed on a band. Thus

is a member of itself, but

is not a member of itself.

In any case, the relation of self-membership that seems to depend on a kind of "pointing to itself" does not require this mode of reference. In the form of a twisted band, the knot-set lives comfortably in three dimensions un-encumbered by decompositions into pointing parts.

The theory of knots and links can be regarded as an extension of a non-standard set theory into the topological domain. This set theory has self-membership and mutuality embedded in it from the very beginning. Viewed from the knot-sets, the theory of knots and links takes its place in the panoply of mathematical systems that embody circularity at different levels of scrutiny. A knot can be regarded as the static expression of a recursive form such as

$$\left\{ \uparrow \right\} = \left\{ \left\{ \uparrow \right\} \right\} = \left\{ \left\{ \left\{ \uparrow \right\} \right\} \right\}$$
$$= \cdots = \left\{ \left\{ \{ \cdots \} \right\} \right\}$$

where here the circularity implicates a structure of recursion and instructions for building an infinite form. It is often thought that a set theory that involves self-membership necessarily involves infinite regress as in

$$\mathcal{M} = \{\mathcal{M}\} \Rightarrow \mathcal{M} = \{\{\mathcal{M}\}\}$$

$$\Rightarrow \mathcal{M} = \{\{\{\mathcal{M}\}\}\}$$

$$\Rightarrow \quad \bullet \bullet \bullet$$

The model of knot-sets and their topological interpretation shows that not only can we give up the idea of a necessary infinite descent in a set that is a member of itself, but we can *also* give up the idea that such a set is fixed in the form of its reference (formally, that it is a fixed point of the operation of braces: {M} = M). The self reference arises in the planar projection.

This brings to a close our discussion of knot-epistemology.

Its themes have echoes in more complex structures.

REFERENCES

1. Y. Aharonov and D. Bohm. Significance of electromagnetic potentials in quantum theory. Phys. Rev. 115. (1959). pp. 485-491.

2. Y. Aharonov and L. Susskind. Observability of the sign change of spinors under 2π rotations. Phys. Rev. Vol. 158. No.5. June 1967, p. 1237.

3. Y.Akutsu and M.Wadati. Knots,links,braids and exactly solvable models in statistical mechanics. Comm.Math.Phys. 117 (1988) 243-259.

4. J.W.Alexander. Topological invariants of knots and links.Trans.Amer.Math.Soc. 20 (1923) .275-306.

5. A. Ashtekar. New perspectives in canonical gravity. Bibliopolis (1988).

6. M.F. Atiyah. **K-Theory.** W.A.Benjamin Inc. (1967).

7. M.F. Atiyah. **Geometry of Yang-Mills Fields.** Accademia Nazionale dei Lincei Scuola Normale Superiore - Lezioni Fermiane. Pisa (1979).

8. M.F. Atiyah. **The Geometry and Physics of Knots.** Cambridge Univ. Press. (1990).

9. S. Axelrod and I. Singer. (to appear)

10. H.J. Bernstein and A.V. Phillips. Fiber bundles and quantum theory. Scientific American. Vol. 245. No.1. July 1981, pp. 122-137.

11. R.J.Baxter. **Exactly Solved Models in Statistical Mechanics** Academic Press (1982).

12. R.J.Baxter. Chromatic polynomials of large triangular lattices. J.Phys.A.:Math.Gen., **20**, 5241-5261.

13. D. Bar-Natan. Perturbative Chern-Simons Theory. (preprint 1990).

14. D. Bar-Natan, Coefficients of Feynman diagrams and Vassiliev's knot invariants. (preprint 1991).

15. E.P. Battey-Pratt and T.J. Racey. Geometric Model for Fundamental Particles. Int. J. Theo. Phys. Vol. 19. No. 6 (1980)

16. B. F. Bayman. Theory of Hitches. Amer. J. Physics. Vol.145., No. 2. (1977), pp. 185-190.

17. J.S.Birman. **Braids, Links and Mapping Class Groups.** Annals of Math. Study No. 82. Princeton University Press (1974).

18. J.S. Birman. On the Jones polynomial of closed 3-braids. Invent. Math. 81. (1985). pp. 287-294.

19. J. S. Birman and X.S. Lin. Knot Polynomials and Vassiliev's Invariants. (preprint 1991).

20. J.S. Birman and R.F. Williams. Knotted Periodic Orbits in Dynamical Systems-I , Lorenz's Equations. Topology. Vol.22. No.1. (1983), pp. 47-82.

21. R.D.Brandt,W.B.R.Lickorish,K.C.Millett. A polynomial invariant for unoriented knots and links. Invent.Math. 84 (1986) 563-573.

22. G.Burde and H.Zieschang. **Knots** . deGruyter(1986).

23. P.N.Burgoyne. Remarks on the combinatorial approach to the Ising problem. J.Math.Phys.Vol.4,No.10,Oct.1963.

24. S.E. Cappell, R. Lee, E.Y. Miller. Invariants of 3-manifolds from conformal field theory. (preprint 1990).

25. J. S. Carter and M. Saito. Reidemeister moves for surface isotopies and their interpretation as moves for movies. (preprint 1991).

26. A. Connes and J. Lott. Particle models and non-commutative geometry. (IHES preprint 1990).

27. Casson,A and C.M.Gordon. On slice knots in dimension three. In **Geometric Topology,** R.J. Milgram ed., pp 39-53, Proc. Symp. Pure Math. XXXII, Amer. Math. Soc. , Providence, R.I.

28. T.P. Cheng and L.F. Li. Gauge theory of elementary particle physics. Clarendon Press - Oxford (1988).

29. G.F. Chew and V. Poenaru. Single-surface basis for topological particle theory. Phys. Rev. D. Vol. 32. No. 10. pp. 2683-2697.

30. J.H.Conway.An enumeration of knots and links and some of their algebraic properties. **Computational Problems in Abstract Algebra** . Pergamon Press,New York(1970),329-358.

31. P. Cotta-Ramusino, E.Guadagnini, M. Martellini and M.Mintchev. Quantum field theory and link invariants. (preprint 1989).

32. L. Crane. Topology of three manifolds and conformal field theories (preprint 1989).

33. L.Crane. 2-D Physics and 3-D Topology. Commun. Math. Phys. 135 (1991), pp. 615-640.

34. L. Crane. Conformal Field Theory, Spin Geometry and Quantum Gravity. (preprint 1990).

35. R.H.Crowell and R.H.Fox. **Introduction to Knot Theory** .Blaisdell Pub. Co. (1963).

36. P.A.M. Dirac. **The Principles of Quantum Mechanics.** Oxford Univ. Press (1958).

37. V.G.Drinfeld. Quantum Groups, Proc.Intl.Congress Math.,Berkeley,Calif.USA(1986).789-820.

38. V.G.Drinfeld. Hopf algebras and the quantum Yang-Baxter equation. Soviet Malth. Dokl. Vol. 32 (1985) No.1.

39. L.D.Faddeev,N.Yu.Reshetikhin,L.A.Takhtajan. Quantization of Lie groups and Lie algebras. LOMI Preprint E-14-87, Steklov Mathematical Institute, Leningrad, USSR.

40. R.A. Fenn and C.P. Rourke. On Kirby's calculus of links. Topology 18 (1979). pp. 1-15.

41. R. P. Feynman. **Theory of Fundamental Processes**. W.A. Benjamin Pub. (1961).

42. D. Finkelstein. Kinks. J. Math. Phys. 7(1966). pp. 1218-1225.

43. D. Finkelstein. Space-Time Code. Phys. Rev. 184 (1969). pp. 1261-1271.

44. D. Finkelstein and J. Rubenstein. Connection between spin, statistics and kinks. J. Math. Pyhs. 9(1968). 1762-1779.

45. R.H. Fox and J.W. Milnor. Singularities of 2-spheres in 4- space and cobordism of knots. Osaka J. Math., **3**, 257-267.

46. R.H. Fox. A quick trip through knot theory. In Topology of 3-manifolds. ed. by M.K. Fort Jr. . Prentice Hall (1962). pp. 120-167.

47. G. Francis. **A Topologists Picturebook.** Springer-Verlag Pub. (1987).

48. D.S. Freed and R.E. Gompf. Computer calculation of Witten's 3-Manifold Invariant. Physical Review Lett. Vol. 66, No. 10. March 1991, pp. 1255-1258.

49. P.Freyd,D.Yetter,J.Hoste,W.B.R.Lickorish,K.C.Millett,A.Ocneanu .A new polynomial invariant of knots and links. Bull.Amer.Math.Soc. 12 (1985) 239-246.

50. D. Friedan, Z. Qiu, S. Shenker. Conformal invariance, unitarity and two-dimensional critical exponents. in **Vertex operators in Mathematics and Physics.** ed by J. Lepowsky, S. Mandelstam, I.M. Singer. pp. 419-450.

51. J.Frohlich. Statistics of fields, the Yang-Baxter equation and the theory of knots and links. (preprint 1987).

52. B. Fuller. Decomposition of the linking number of a closed ribbon: a problem from molecular biology. Proc. Natl. Acad. Sci. USA, Vol. 75. No.8. (1978). pp. 3557-3561.

53. K. Gawedski. Conformal Field Theory. Sem. Bourbaki. 41e annee (1988-89). 704. pp. 1-31.

54. M.L. Glasser. Exact partition function for the two-dimensional Ising model. Amer. J. Phys. Vol. 38. No. 8. August 1970. pp. 1033-1036.

55. D.L. Goldsmith. The Theory of Motion Groups. Michigan Math. J. 28. (1981). pp. 3-17.

56. D.L. Goldsmith. Motion of Links in the Three Sphere. Math. Scand. 50 (1982). pp. 167-205.

57. B. Grossman. Topological quantum field theories: relations between knot theory and four manifold theory. (preprint 1989).

58. E. Guadagnini, M. Martellini, M. Mintchev. Perturbative aspects of the Chern-Simons field theory. Phys. Lett. B227 (1989). p 111.

59. P.de la Harpe, M.Kervaire, and C.Weber. On the Jones polynomial. L'Enseign.Math. 32 (1986) 271-335.

60. R.Hartley. Conway potential functions for links, Comment.Math.Helv. 58 (1983) 365-378.

61. B. Hasslacher and M. J. Perry. Spin networks are simplicial quantum gravity. Physics Letters, Vol. 103B, No. 1, July 1981.

 62. P.J. Heawood. On the four-colour map theorem. Quart. J. Math., vol. 29, pp. 270-285 (1898).

63. M. A. Hennings. A polynomial invariant for oriented banded links. (preprint 1989).

64. M. A. Hennings. A polynomial invariant for unoriented banded links. (preprint 1989).

65. M. A. Hennings. Hopf algebras and regular isotopy invariants for link diagrams. (to appear in Math. Proc. Cambridge Phil. Soc.)

66. M. A. Hennings. Invariants of links and 3-manifolds obtained from Hopf algebras. (to appear).

67. C.F.Ho. A new polynomial invariant for knots and links - preliminary report. AMS Abstracts, Vol.6 No.4,Issue 39(1985) p. 300.

68. J.Hoste. A polynomial invariant of knots and links. Pacific J. Math. 124 (1986) 295-320.

69. F.Jaeger. A combinatorial model for the Homfly polynomial. (preprint 1988).

70. F.Jaeger. Composition products and models for the Homfly polynomial. L'Enseignment Math. t35 (1989). pp 323-361.

71. F. Jaeger. Tutte polynomials and link polynomials. Proc. Amer. Math. Soc. 103 (1988). pp. 647-654.

72. F. Jaeger. On edge colorings of cubic graphs and a formula of Roger Penrose. Ann. of Discrete math. 41 (1989). pp. 267-280.

73. F. Jaeger, D.L. Vertigan and D.J.A Welsh. On the computational complexity of the Jones and Tutte polynomials. Math. Proc. Camb. Phil. Soc. Vol. 108 (1990). pp. 35-53.

74. F. Jaeger, L.H. Kauffman and H. Saleur. (in preparation)

75. H. Jehle. Flux quantization and particle physics. Phys. Rev. D. Vol. 6. No.2. (1972). pp. 441-457.

76. M.Jimbo. A q-difference analogue of U(q) and the Yang-Baxter equation. Lect. in Math. Physics 10 (1985) 63-69.

77. M.Jimbo. Quantum R-matrix for the generalized Toda system. Comm.Math.Phys. 102 (1986) 537-547

78. V.F.R.Jones. A new knot polynomial and von Neumann algebras. Notices of AMS 33 (1986) 219-225.

79. V.F.R.Jones.A polynomial invariant for links via von Neumann algebras. Bull.Amer.Math.Soc. 129 (1985) 103-112.

80. V.F.R.Jones. Hecke algebra representations of braid groups and link polynomials.Ann. of Math. 126 (1987) 335-388.

81. V.F.R.Jones.On knot invariants related to some statistical mechanics models. Pacific J. Math. , Vol. 137, No.2 (1989), 311-334.

82. V.F.R. Jones. On a certain value of the Kauffman polynomial. (to appear in Comm. Math. Phys.).

83. V.F.R. Jones. Subfactors and related topics. In Operator algebras and Applications Vol. 2. ed. by D.E. Evans and M. Takesaki. Cambridge Univ. Press (1988). pp. 103-118.

84. V.F.R. Jones. Index for subfactors. Invent. Math. 72 (1983). pp 1-25.

85. D. Joyce. A classifying invariant of knots, the knot quandle. J. Pure Appl. Alg., 23, 37-65.

86. L.H.Kauffman.The Conway polynomial.Topology 20 (1980) 101-108.

87. L.H.Kauffman. **Formal Knot Theory** . Princeton University Press Mathematical Notes #30 (1983).

88. L.H.Kauffman. **On Knots**. Annals of Mathematics Study 115, Princeton University Press (1987).

89. L.H.Kauffman. State Models and the Jones Polynomial. Topology 26 (1987) 395-407.

90. L.H.Kauffman.Invariants of graphs in three-space. Trans.Amer.Math.Soc.Vol.311 2 (Feb. 1989) 697-710.

91. L.H.Kauffman. New invariants in the theory of knots. (lectures given in Rome,June 1986.) - Asterisque 163-164 (1988),p.137-219.

92. L.H.Kauffman. New invariants in the theory of knots. Amer. Math. Monthly Vol.95,No.3,March 1988. pp 195-242.

93. L.H.Kauffman. An invariant of regular isotopy. Trans. Amer. Math. Soc. Vol. 318. No. 2 (1990). pp. 417-471.

94. L.H.Kauffman. Statistical Mechanics and the Alexander Polynomial. Contemp. Math. Vol. 96 (1989). Amer. Math. Soc. Pub. pp. 221-231.

95. L.H.Kauffman.Statistical mechanics and the Jones polynomial. In Proceedings of the 1986 Santa Cruz conference on Artin's Braid Group. AMS Contemp. Math. Series. Vol. 78. (1989). pp. 263-297. reprinted in **New Problems, Methods and Techniques in Quantum Field Theory and Statistical Mechanics**. pp. 175-222. ed. by M. Rasetti. World Scientific Pub. (1990).

96. L.H.Kauffman. A Tutte polynomial for signed graphs. Discrete Applied Math. Vol. 25. (1989). pp. 105-127.

97. L.H.Kauffman. State models for link polynomials. l'Enseignment Math. t.36 (1990). pp. 1-37.

98. L.H.Kauffman. Polynomial Invariants in Knot Theory. **Braid Group, Knot Theory and Statistical Mechanics**. Ed. C.N.Yang and M.L.Ge. World Sci. Pub. Advanced Series in Mathematical Physics, Vol. 9. (1989), 27-58.

99. L.H.Kauffman. Spin networks and knot polynomials. Intl. J. Mod. Phys. A. Vol. 5. No. 1. (1990). pp. 93-115.

100. L.H. Kauffman. Transformations in Special Relativity. Int. J. Theo. Phys. Vol. 24. No. 3. pp. 223-236. March 1985.

101. L. H. Kauffman and H. Saleur. Free fermions and the Alexander-Conway polynomial. (to appear in Comm. Math. Phys.).

102. L. H. Kauffman and H. Saleur. Fermions and link invariants. (preprint 1991).

103. L.H. Kauffman and H. Saleur. Map coloring and the Temperley-Lieb Algebra . (In preparation.)

104. L.H. Kauffman. Gauss Codes and Quantum Groups. (preprint 1991)

105. L. H. Kauffman and D.E. Radford. A necessary and sufficient condition for a finite-dimensional Drinfeld double to be a ribbon Hopf algebra. (preprint 1991).

106. L. H. Kauffman. Knots, Spin Networks and 3-Manifold Invariants. (to appear in the proceedings of KNOTS 90 - Osaka Conference).

107. L.H. Kauffman. **Knots and Physics**, World Scientific Pub. (1991).

108. L.H.Kauffman and Sostenes Lins. Computing Turaev-Viro Invariants for 3-Manifolds. Manuscripta Math. 72, pp. 81-94 (1991).

109. L. H. Kauffman and S. Lins. A 3-manifold invariant by state summation. (preprint 1990).

110. L. H. Kauffman. Knots, abstract tensors, and the Yang-Baxter equation. In **Knots, Topology and Quantum Field theories** - Proceedings of the Johns hopkins Workshop on Current Problems in Particle Theory 13. Florence (1989). ed. by L. Lussana. World Scientific Pub. (1989). pp. 179-334.

111. L. H. Kauffman and P. Vogel. Link polynomials and a graphical calculus. (announcement 1987. preprint 1991).

112. L.H. Kauffman. Problems in Knot Theory - Chapter in book **Open Problems in Topology** (ed. by J. van Mill and pub. by North Holland 1990).

113. L. H. Kauffman. Map coloring and the vector cross product. J. Comb. Theo. Ser. B. Vol. 48. No. 2. April 1990. pp. 145- 154.

114. L.H. Kauffman. From knots to quantum groups and back. In proceedings of the fall 1989 Montreal conference on Hamiltonian systems. ed. by Harnad and Marsden, CRM Pub. (1990), pp. 161-176. In expanded form in **Quantum Groups** ed. by Curtright, Fairlie and Zachos, World Sci. (1991), pp. 1-32.

115. C. Anezeris, A.P. Balachandran, L. Kauffman, A.M. Srivastava. Novel statistics for strings and string 'Chern Simons' terms. (preprint 1990).

116. L. H. Kauffman. An integral heuristic. Intl. J. Mod. Phys. A. Vol. 5. No. 7. (1990). pp. 1363-1367.

117. L.H. Kauffman. Special relativity and a calculus of distinctions. In **Proceedings of the 9th Annual International Meeting of the Alternative Natural Philosophy Association - Cambridge University, Cambridge, England (September 23, 1987).** Published by ANPA West, Palo Alto, Calif. pp. 290-311.

118. T. Kanenobu. Infinitely many knots with the same polynomial. Proc. Amer. Math. Soc. 97 (1986). pp. 158-161.

119.   A. B. Kempe. On the geographical problem of the four colors. Amer. J. Math. Vol. 2. (1879). pp. 193-200.

120.   R. Kirby. A calculus for framed links in S^3. Invent. Math., 45, 35-36. (1978)

121.   R. Kirby and P. Melvin. On the 3-manifold invariants of Reshetikhin- Turaev for sl(2,C). (preprint 1990).

122.   T.P.Kirkman.The enumeration,description and construction of knots with fewer than 10 crossings. Trans. Royal Soc. Edin. 32(1865),281-309.

123.   A.N. Kirillov. and N.Y. Reshetikhin. Representations of the algebra $U_q(sl_2)$, q-orthogonal polynomials and invariants of links. In Infinite dimensional Lie algebras and groups. ed. by V.G. Kac. Adv. Ser. in Math. Phys. Vol. 7. (1988). pp. 285-338.

 124.   T.Kohno. Monodromy representations of braid groups and Yang-Baxter equations. Ann.Inst.Fourier,Grenoble 37 ,4 (1987) 139-160.

125.   T. Kohno. Topological invariants for 3-manifolds using representations of mapping class groups I. (preprint 1990).

126.   M. Kontsevich. Rational conformal field theory and invaraints of 3-dimensional manifolds. (preprint 1988).

127.   P.P.Kulish,N.Yu.Reshetikhin and E.K.Sklyanin. Yang-Baxter equation and representation theory:I. Letters in Math. Physics 5 (1981) 393-403.

128.   P.P.Kulish and E.K. Sklyanin. Solutions of the Yang-Baxter equation. J. Soviet Math. 19 (1982) pp. 1596-1620. {reprinted in **Yang-Baxter Equation in Integrable Systems.** ed. by M. Jimbo. World Scientific Pub. (1990)}

129.   G. Kuperberg. Involutory Hopf algebras and 3-manifold invariants. (preprint 1990).

130.   G. Kuperberg. The quantum G_2 link invariant. (preprint 1990).

131. B.I. Kurpita and K.Murasugi. On a heirarchical Jones invariant. (preprint 1990).

132. L.A. Lambe and D. E. Radford. Algebraic aspects of the quantum Yang-Baxter equation. (preprint 1990).

133. R.G. Larson and J. Towber. Braided Hopf algebras: a dual concept to quasitriangularity. (preprint 1990).

134. R.Lawrence. A universal link invariant using quantum groups. (preprint 1988).

135.   R. Lawrence. A functorial approach tto the one-variable Jones polynomial. (preprint 1990).

136. H.C. Lee. Q-deformation of sl(2,C) x ZN and link invariants. (preprint 1988).

137. H. C. Lee. Tangles, links and twisted quantum groups. (preprint 1989).

138. H.C.Lee,M.Couture and N.C.Schmeling, Connected link polynomials (preprint 1988).

139. W.B.R.Lickorish and K.C.Millett. A polynomial invariant for oriented links. Topology 26 (1987) 107-141.

140. W.B.R.Lickorish and K.C.Millett. The reversing result for the Jones polynomial. Pacific J. Math. 124. (1986). pp. 173-176.

141. W.B.R. Lickorish. A representation of orientable, combinatorial 3-manifolds. Ann. Math. 76 (1962). pp. 531-540.

142. W.B.R.Lickorish. Polynomials for links. Bull. London Math. Soc., 20, 558-588. (1988).

143. W. B. R. Lickorish. 3-Manifolds and the Temperley Lieb Algebra. (preprint 1990).

144. W.B.R. Lickorish. Calculations with the Temperley-Lieb algebra. (preprint 1990).

145. C.N.Little. Non-alternate +- knots. Trans. Royal Soc. Edin. 35 (1889) 663-664.

146. S. Maclane. **Categories for the Working Mathematician.** Springer-Verlag (1971).

147. J.M. Maillet and F.W. Nijhoff. Multidimensional integrable lattices, quantum groups and the D-simplex equations. (preprint - Institute for Non-Linear Studies - Clarkson University - INS #131 (1989)).

148. S.Majid. Quasitriangular Hopf algebras and Yang-Baxter equations. Intl. J. Mod. Phys. A. Vol.5. No. 1. (1990). pp. 1-91.

149. Yu.I.Manin. Quantum groups and non-commutative geometry. (Lecture Notes, Montreal, Canada , July 1988).

150. P.P. Martin. Analytic properties of the partition function for statistical mechanical models. J. Phys. A: Math. Gen. 19 (1986). pp. 3267-3277.

151. S.V. Matveev. Transformations of special spines and the Zeeman conjecture. Math. USSR izvestia 31: 2 (1988). pp. 423-434.

152. D.A. Meyer. State models for link invariants from the classical Lie groups. (preprint 1990).

153. E. W. Mielke. Knot wormholes in geometrodynamics. General Relativity and Gravitation. Vol. 8. No. 3. (1977). pp. 175-196.

154. J. W. Milnor. **Singular Points of Complex Hypersurfaces.** Princeton Universtity Press (1968).

155. G. Moore and N. Seiberg. Classical and quantum conformal field theory. Commun. Math. Phys. 123 (1989). pp. 177-254.

156. H. R. Morton. The Jones polynomial for unoriented links. Quart. J. Math. Oxford (2). 37. (1986). pp. 55-60.

157. H.R.Morton and H.B. Short. The two-variable polynomial for cable knots. Math. Proc. Camb. Phil. Soc. 101. (1987). pp. 267-278.

158. H.R. Morton and P.M. Strickland. Satellites and surgery invariants. (preprint 1990).

159. J. P. Moussouris. Quantum models of space-time based on recoupling theory. (Mathematics Thesis, Oxford University - 1983).

160. J. Murakami. The parallel version of link invariants. preprint Osaka Univ. (1987).

161. J. Murakami. A state model for the multi-variable Alexander polynomial. (preprint 1990).

162. J. Murakami. On local relations to determine the multi-variable Alexander polynomial of colored links. (preprint 1990).

163. K. Murasugi. The Jones polynomial and classical conjectures in knot theory. Topology, 26, 187-194. (1987).

164. K. Murasugi. Jones polynomials and classical conjectures in knot theory II. Math. Proc. Camb. Phil. Soc. 102 (1987). 317-318.

165. M.H.A. Newman. On a string problem of Dirac. J. London Math. Soc. 17 (1942). pp. 173-177.

166. R.Penrose.Applications of negative dimensional tensors. **Combinatorial Mathematics and its Applications.** Edited by D.J.A.Welsh.Academic Press (1971).

167. R. Penrose. Angular momentum: an approach to Combinatorial Space-Time. In **Quantum Theory and Beyond.** ed. T.A. Bastin. Cambridge Univ. Press (1969).

168. R. Penrose. Combinatorial quantum theory and quantized directions. **Advances in Twistor Theory.** ed by L.P. Hughston and R.S. Ward. Pitman (1979). pp. 301-307.

169. J.H.H. Perk and F.Y. Wu. Nonintersecting string model and graphical approach: equivalence with a Potts model. J. Stat. Phys. Vol. 42. Nos. 5/6/ (1986). pp. 727-742.

170. J.H.H.Perk and C.L. Schultz. New families of commuting transfer matrices in q-stqate vertex models. Physics Letters. Vol. 84A. Number 8. August 1981. pp. 407-410.

171. A.M. Polyakov. Fermi-Bose transmutations induced by gauge fields. Mod. Phys. Lett. A3 (1987). pp. 325-328.

172. G. Ponzano and T. Regge. Semiclassical limit of Racah coefficients. in **Spectroscopic and Group Theoretical Methods in Physics.** (ed. F. Bloch) North- Holland Publ. Co., Amsterdam (1968).

173. J.H. Przytycki and P. Traczyk. Invariants of links of Conway type. Kobe J. Math. 4 (1987). pp. 115-139.

174. M. Rasetti and T. Regge. Vortices in He II, Current Algebras and Quantum Knots. Physica 80a (1975). pp. 217-233. Quantum vortices. in **Highlights of Condensed Matter Theory** (1985). Soc. Italiana di Fisica - Bologna - Italy.

175. K.Reidemeister. **Knotentheorie.** Chelsea Publishing Co.,New York (1948) Copyright 1932. Julius Springer,Berlin.

176. N.Y.Reshetikhin. Quantized universal enveloping algebras, the Yang-Baxter equation and invariants of links, I and II. LOMI reprints E-4-87 and E-17-87, Steklov Institute, Leningrad, USSR.

177. N. Yu. Reshetikhin. Invariants of links and 3-manifolds related to quantum groups. (preprint 1990).

178. N.Y. Reshetikhin and V. Turaev. Ribbon graphs and their invariants derived from quantum groups. Comm. Math. Phys. 127 (1990). pp. 1-26.

179. N.Y. Reshetikhin and V. Turaev. Invariants of Three Manifolds via link polynomials and quantum groups. (preprint 1989. to appear in Invent. Math.).

180. G.D. Robertson. Torus knots are rigid string instantons. Univ. of Durham - Centre for Particle Theory. Preprint (1989).

181. D.Rolfsen. **Knots and Links.** Publish or Perish Press (1976).

182. M.Rosso. Groupes quantiques de Drinfeld et Woronowicz. (preprint 1988).

183. M. Rosso. Groupes quantiques et modeles a vertex de V. Jones en theorie des noeuds. C.R.Scad.Sci.Paris t.307 (1988). pp. 207-210.

184. C.Rovelli and L.Smolin. Knot theory and quantum gravity. (preprint 1988).

185. T. L. Saaty and P.C. Kainen. **The Four Color Problem.** Dover Pub. (1986).

186.   H. Saleur. Virasoro and Temperley Lieb algebras. In **Knots, Topology and Quantum Field theories** - Proceedings of the Johns hopkins Workshop on Current Problems in Particle Theory 13. Florence (1989). ed. by L. Lussana. World Scientific Pub. (1989). pp. 485-496.

187.   H. Saleur. Zeroes of chromatic polynomials: a new approach to Beraha conjecture using quantum groups. Commun. Math. Phys. 132 (1990). pp. 657-679.

188.   S. Samuel. The use of anticommuting variable integrals in statistical mechanics. I. The computation of partition functions. II. The computation of correlation functions. III. Unsolved models.   J. Math. Phys. 21(12) (1980). pp. 2806-2814, 2815-2819, 2820- 2833.

189.   G. Segal. Two-dimensional conformal field theories and modular functors. **IXth International Congress on Mathematical Physics - 17-27 July-1988-Swansea, Wales**, Ed. Simon, Truman, Davies, Adam Hilger Pub. (1989). pp. 22-37.

190.   E.K. Skylanin. Some algebraic structures connected with the Yang-Baxter equation. Representation of quantum algebras. Funct. Anal. Appl. 17 (1983). pp. 273-284.

191.   L.Smolin. Link polynomials and critical points of the Chern-Simon path integrals. Mod. Phys. Lett. A. Vol. 4. No. 12 (1989). pp. 1091-1112.

192.   L. Smolin. Quantum gravity in the self-dual representation. Contemp. Math. Vol. 28 (1988).

193.   R. Sorkin. Particle statistics in three dimensions. Phys. Rev. D27 (1983). pp 1787-1797.

194,   R. Sorkin. A general relation between kink exchange and kink rotation. Commun. Math. Phys. 115 (1988). pp. 421-434.

195.   P.G.Tait.On Knots I,II,III.Scientific Papers Vol.I,Cambridge University Press.,London,1898,273-347.

196.   H.N.V. Temperley and E.H.Lieb. Relations between the 'percolation' and 'coloring' problem and other graph-theoretical problems associated with regular planar lattices: some exact results for the 'percolation' problem. Proc. Roy. Soc. Lond. A 322. (1971). pp. 251-280.

197.   M.B.Thistlethwaite. Knot tabulations and related topics. **Aspects of Topology.** Ed. I.M.James and E.H.Kronheimer, Cambridge University Press (1985)1-76.

198.   M.B.Thistlethwaite. A spanning tree expansion of the Jones polynomial. Topology 26 (1987). pp. 297-309.

199. M.B.Thistlethwaite. On the Kauffman polynomial of an adequate link. Invent. Math. 93. (1988). pp. 285-296.

200. M.B. Thistlethwaite. Kauffman's polynomial and alternating links. Topology 27 (1988). pp. 311-318.

201. W. Thompson (Lord Kelvin) . On vortex atoms. Philosophical magazine. 34. July 1867. pp. 15-24. **Mathematical and Physical Papers,** Vol. 4. Cambridge (1910).

202. B.Trace. On the Reidemeister moves of a classical knot. Proc.Amer.Math.Soc. Vol.89,No.4 (1983).

203. V.G.Turaev. The Yang-Baxter equations and invariants of links. LOMI preprint E-3-87, Steklov Institute, Leningrad, USSR. Inventiones Math. 92 Fasc.3,527-553.

204. V.G. Turaev and O. Viro. State sum invariants of 3-manifolds and quantum 6j symbols. (preprint 1990).

205. V. G. Turaev. Quantum invariants of links and 3-valent graphs in 3-manifolds. (preprint 1990).

206. V.G. Turaev. Shadow links and face models of statistical mechanics. Publ. Inst. Recherche Math. Avance. Strasbourg (1990).

207.   V.G. Turaev. Quantum invariants of 3-manifolds and a glimpse of shadow topology (preprint 1990).

208. W.T. Tutte. Graph Theory. Encyclopedia of Mathematics and its Applications 21 . Cambridge University Press. (1984).

209. V.A. Vassiliev, Cohomology of knot spaces. in **Theory of Singularities and Its Applications** (ed. V.I. Arnold), Advances in Soviet Math., Vol.1, AMS (1990).

210. K. Walker. On Witten's 3-Manifold Invariants. (preprint 1991).

211. H. Wenzl. Representations of braid groups and the quantum Yang-Baxter equation. Pacific J. Math. 145 (1990). pp. 153-180.

212. H. Weyl. Gravitation und Elektrizitat. Sitzungsberichte der Koniglich Preussischen Akademie der Wissenshaften. 26 (1918). pp. 465-480.

213. H.Whitney. On regular closed curves in the plane. Comp.Math. 4 (1937),276-284.

214. H. Whitney. A logical expansion in mathematics. Bull. Amer. Math. Soc., 38, 572-579.

215. E.Witten. Quantum field theory and the Jones polynomial. Commun.Math.Phys. <u>121</u> , 351-399 (1989).

216. E.Witten. Gauge Theories vertex models and quantum groups. Nucl. Phys. B. 330. (1990). pp. 225-346.

217. E. Witten. Global gravitational anomalies. Comm. Math. Phys. 100. (1985). pp. 197-229.

218. S. Winker. **Quandles, Knot Invariants and the N-fold Branched Cover.** Ph.D. Thesis, University of Illinos at Chicago (1984).

219. F.Y. Wu. The Potts Model. Rev. Mod. Phys. Vol. 54. No. 1. Jan. 1982. pp. 235-268.

220. F.Y.Wu. Graph theory in statistical physics. in **Studies in Foundations and Combinatorics**-Advances in Mathematics Supplementary Studies. Vol.1. (1978). Academic Press Inc. pp. 151-166.

221. D. Yetter. Quantum groups and representations of monoidal categories. Math. Proc. Camb. Phil. Soc. 108 (1990). pp. 197-229.

222. M. Zaganescu. Bosonic knotted strings and cobordism theory. Europhys. Lett. 4 (5) (1987) . pp. 521-525.

223. A.B.Zamolodchikov. Factorized S matrices and lattice statistical systems. Soviet Sci. Reviews. PartA (1979-1980).

224. A.B.Zamolodchikov. Tetrahedron equations and the relativistic S-matrix of straight strings in 2+1 dimensions. Comm.Math.Phys. <u>79</u> (1981)489-505.

225. A.B.Zamolodchikov. Factorized S-matrices and lattice statistical systems. Soviet Sci. Reviews Part A. (1979).

Department of Mathematics, Statistics and Computer Science (M/C 249)
The University of Illinois at Chicago (Box 4348)
Chicago, Illinois 60680
e-mail: U10451@UICVM.BITNET

Index

achiral, 45
Alexander polynomial, 131, 132, 134, 211, 225
Alexander-Conway polynomial, 225
algebraic topology, 131
almost unknotted embedding, 82, 83
almost unknotted graphs, 75
alternating diagram, 46
ambient isotopy, 42, 156
amplitude, 140, 141, 143, 144
angular momentum, 195
angular momentum recoupling, 177
annihilation of particles, 143
annihilations, 141
antisymmetrizer, 204
Artin braid group, 131, 139, 161
atoms, 131

base pairs, 54
BFACF algorithm, 88
binor calculus, 201
bra-ket, 142
bracket evaluation, 163
bracket identity, 173
bracket polynomial, 153, 157, 169, 198
braids, 161
bras, 142
Burau representation, 215

cap, 187
categories, 145
catenane, 2, 59, 99, 104
Chebyshev polynomial, 176
Chern-Simons Lagrangian, 140, 222
chiral, 45, 100
chirality, 99
chromatic polynomial, 153, 170
classifying vector, 47
collapse of the wave function, 144
combinatorial knot theory, 186
combinatorial state model, 219
combinatorics, 133, 153
completeness, 143
complex numbers, 140
composite, 45
composition of morphisms, 188

conceit, 137
conformal field theory, 133, 139, 140, 186, 224
conformal quantum field theory, 172
connective constant, 77, 78
Conway, 132
creations, 141
cross-channeling, 134
crossing, 135, 143, 187, 191
crossing number, 45, 81
cup, 187
curvature, 223
curvature tensor F, 222

determinant, 211
Dirac formalism, 143
Dirac string trick, 146
direct repeats, 58
distributive recombination, 59
DNA on protein complexes, 25
DNA repair, 6
DNA replication, 4
DNA topology, 1
DNAse, 54
Dubrovnik polynomial, 198

elementary braid, 161
Elliot-Biedenharn identity, 180, 181
energy, 168
entanglement complexity, 75, 81, 89, 91
entanglement number, 90
enzyme mechanism, 61
epsilon, 194
ether, 131
evolution of the topological structure of DNA, 1

Feynman integrals, 133
Feynman path integral, 137, 141
Four Color Theorem, 170, 172
4-plat, 47
4-valent plane graph, 154
Fox derivation, 214
framed link, 220
framing, 226, 227
free differential calculus, 211, 213, 215

Recent Titles in This Series

(*Continued from the front of this publication*)

(See the AMS catalog for earlier titles)